How to pass examinations in

STATISTICS

D.V. Friend
M.Sc., M.Sc. (Econ.), F.I.S.
Principal Lecturer in Statistics
Polytechnic of Central London

Cassell • London

Cassell Ltd.
35 Red Lion Square, London WC1R 4SG
and at Sydney, Auckland, Toronto, Johannesburg,
an affiliate of Macmillan Publishing Co. Inc.,
New York

First published 1980

I.S.B.N. 0 304 30558 8

Reproduced from copy supplied
printed and bound in Great Britain
by Billing & Sons Limited
Guildford, London, Oxford, Worcester

Contents

Introduction

This book is based on notes given to students in the School of Social Sciences and Business Studies of the Polytechnic of Central London. The notes first appeared in 1968 and have been revised on several occasions. I have had extensive experience as a teacher, examiner and assessor in statistics and over the years I have tried and tested, with considerable success, the teaching material on which this book is based. Many of my students who have passed an examination in statistics have not been very numerate, so that the material presented should be readily understood even by the weakest candidates. My colleagues and the students have made many suggestions for improvements and, where appropriate, these suggestions have been incorporated in the notes and in consequence in this book. I would, therefore, like to take this opportunity of thanking my colleagues and the students for the encouragement they have given me.

I have over twenty years' experience as a teacher, and I have been on the staff of a grammar school, a technical college and a polytechnic. I have taught mathematics and statistics to students with a very wide range of mathematical ability, but I have specialised in teaching the weak students. I have acted as an examiner in statistics for London University, the Institute of Statisticians and the Royal Society of Arts. I have been an Assessor in Statistics for the Business Education Council for Ordinary and Higher National Certificates and Diplomas in Business Studies. This has given me the opportunity to see the effectiveness of various teaching methods used in institutions in the United Kingdom and overseas. I have noticed that some institutions, when teaching a topic such as regression, use a method of calculation whereby only the better students obtain the correct result; other institutions use a different method of calculation and students attempting the same type of problem are much more successful. The notes used at the Polytechnic of Central London, and thus the contents of this book, have incorporated the most successful methods observed in colleges and polytechnics throughout the country.

Most elementary books on statistics do not include solutions to the exercises. A novel feature of this book is that solutions to all the exercises are provided and answers are given to the descriptive parts of questions.

It is sometimes argued that possession of 'O' level mathematics is essential for success in statistics. In my view, whilst 'O' level mathematics is desirable it is not essential. For the past three years I have taught statistics on an evening degree course for B.A. social science; 'O' level mathematics is not a requirement for this course and about a third of the students do not have it. Nevertheless, in this three year period, there have been 114 candidates and 110 have passed, showing quite clearly that lack of 'O' level mathematics has not proved a handicap. This book has been specifically designed to help the student, particularly the student with limited numeracy, to pass an examination in statistics with minimum pain.

The book does not cover t and F tests. These topics are not often included in the syllabuses for which this book is designed. Students who wish to study these topics and would like a book which has the same format as this book — worked examples, exercises and solutions — are recommended to purchase *Statistical Methods in Management*, volume 1, by Tom Cass, published by Cassell.

This book covers most of the usual areas for statistics examinations set in the social science and business studies fields and by the professional bodies. There are, however, no chapters on sources of economic, social and business statistics — this is a large area and it is necessary for the student to consult the standard works in this field, *Sources of Economic and Business Statistics* and *Sources of Social Statistics*; the author of both books is Bernard Edwards and the publisher is Heinemann. As an examiner I am fully aware, however, that this topic is very badly neglected by students. Examination questions set on sources of statistics are largely ignored. If you are reasonably numerate and the examination paper does not force you to answer questions in this area, you could probably risk ignoring this topic. On the other hand if you are very innumerate and questions are regularly set on sources of statistics, then you would clearly gain by reading the books mentioned above.

The worked examples and the exercises in my book are almost all taken from past examination questions. In most cases the source of the question is given. In some cases a part of an examination question has been used or an examination question amended; in this situation the source of the question has not been stated. My experience as an assessor has made me aware that examiners occasionally take examination questions from textbooks. If I have, therefore, inadvertently used a question from a textbook, I offer my humble apologies.

Examination Strategy

To pass an examination in statistics requires a plan of action:

1. Find out the sort of questions which are set in the examination.

2. Be certain that you, the candidate, have a reasonable chance of successfully obtaining the correct solution to the popular questions.

It is essential to find out about the examination you are going to take. You need **(i)** the syllabus, and **(ii)** as many past examination papers as you can obtain, particularly the most recent.

The syllabus will tell you the topics you need to cover; whether a formulae list is supplied in the examination room; whether you can take a calculator into the examination room and which tables — statistical, logarithmic etc. — will be supplied. If a formulae list is supplied, try to obtain a copy so that you can see what is included; any formulae not included which may be relevant to examination questions you will have to learn. Try to obtain a copy of the tables provided, since statistical tables do differ in format, and it is essential to know how to use the tables supplied in the examination room, which may well be different from those you have used in your preparation. If a calculator is allowed, I would strongly advise that you purchase one, because some examiners set questions involving more difficult data if calculators are allowed. A cheap calculator is better than no calculator. Buy a calculator well before the examination, preferably at the beginning of the course, so that you can learn to use it efficiently. Most examination boards object to programmable and certain pre-programmed calculators. Make certain that any calculator you plan to use in the examination room is acceptable to the examination authorities well before the examination day. Before the examination (not just five minutes before the examination is due to start), check over the calculator to see that it is working correctly. Take into the examination room a set of spare batteries.

The syllabus tells you the topics covered. The actual examination papers show how the examiner interprets the syllabus. Is the examination paper divided into sections? Is there a compulsory question? How many questions are set? How many questions do you have to attempt? If the paper is sectionalised, how many questions must be answered from each section?

An analysis of the last (say) six examination papers will enable you to draw up a table of the frequency of certain questions. This would, for example, show whether a question on mean and standard deviation is set every year or almost every year. Such a frequency table would give a good guide to the topics which need to be studied in depth; attempt also plenty of exercises on such topics. Those topics which are set infrequently you could perhaps risk covering in less detail. It is also essential to see whether questions on statistical method simply involve a calculation or whether they include, in addition, an interpretative element.

Certain examination boards issue model solutions and/or comments on students' attempts at questions; these are well worth obtaining and studying in detail.

Many examination boards set examination papers which include 'essay-type' questions, e.g. on survey methods and on sources of economic, social and business statistics. Some candidates, who are very innumerate, try to pass by concentrating on such questions; my experience as an examiner is that relatively few of these candidates are successful. *It is essential that all examination candidates should be able to do some of the calculations*. If you are very innumerate, try to ensure that you can make a reasonable attempt at a limited number of topics.

Many examination papers in statistics ask you to answer five questions in three hours. Allowing five minutes to read the paper, this means 35 minutes per question. Your job in the examination room is to maximise your marks. Some students fail simply because they are badly organised in the examination room — they attempt perhaps only 3, $3\frac{1}{2}$ or 4 questions. Always try to attempt 5 questions, allowing 35 minutes for each question (take a watch or a clock with you into the examination room — not all examination rooms have a clock or one that is working). If you have not finished a question at the end of 35 minutes, start the next question. In practice you probably obtain most of your marks on a particular question during the first 20 minutes. Taking an hour to answer one question is simply an inefficient way of using your time.

Many examination questions in statistics involve a calculation and a descriptive element. Usually, but not always, the calculation carries more marks than the descriptive element. Some students prefer to do the calculations of all the questions first and then to return to the descriptive element of the questions later, having left a space at the end of the calculation part of each question for this purpose. Timing of individual questions is not so easy by this method, but if the candidate is relatively speedy in doing the calculations, the descriptive part is often better done simply because the candidate is not in so much of a hurry to proceed to the next question. I do not particularly recommend this approach, but I can report that some students have used it very successfully.

A careful analysis of the requirements of the examination; the correct tools — calculator, spare batteries, ruler, rubber, pencil, pencil sharpener, protractor etc.; familiarity with the style of the paper; sensible use of examination time; these should give you every confidence when you read the paper and start to answer the first question.

How to use this book

With all mathematical subjects success comes by practice — by attempting as many of the exercises as you can in the time at your disposal. Merely reading the text and looking at the worked examples *without* doing the exercises is an almost certain way to fail.

In all chapters a worked example is set on each topic and following each worked example there is an exercise which is similar to the worked example but with different data. At the end of the chapter a miscellaneous set of exercises covers all the topics of the chapter. The best method is to study the text; study the worked example; and then do the exercise that follows the worked example. When you have completed the chapter, attempt the exercises at the end of the chapter. The appendix gives solutions or outline solutions to all the exercises. If you cannot obtain the numerical answer to the question, check your working — you may have made an arithmetical or method error. If you are still unsuccessful, then compare your work with the solution in the appendix in detail and find out exactly where you went wrong; then try a similar question from the exercises at the end of the chapter.

Many examination questions involve a comment or interpretative element in addition to a calculation. The exercises in this book are mostly taken from examination papers and in consequence some include a descriptive part. In designing this book the appendix has been used to cover the typical descriptive element of questions. In my experience students like to have compact answers to descriptive questions rather than attempt to extract the information from the relevant chapter. It is essential, therefore, for you to read the answers in the appendix to the descriptive part of certain exercises. Such exercises are marked by a vertical line in the margin, and the solutions are similarly marked. I accept that some of the descriptive answers to certain questions may offend the statistical purist, but my aim is to give candidates a few sentences of comment that they can reproduce in the examination room and thereby score some marks.

File your solutions to the exercises so that they are readily accessible during the revision period. As examination time approaches you may well have forgotten how to do some of the questions, particularly those from earlier chapters. You must revise systematically. Attempt one or two questions from the exercises at the end of each chapter. If you obtain the correct answer, proceed to the next chapter, but if you have difficulty, you need to go through the worked examples and the exercises for the chapter again, referring to the solutions you obtained when you went through the chapter for the first time.

Remember, success comes by doing exercises and obtaining the correct answer.

Acknowledgements

I gratefully acknowledge permission to use examination questions from the following bodies:

University of London
Institute of Statisticians (I.O.S.)
Royal Society of Arts (R.S.A.)
The Polytechnic of Central London

The majority of the questions have been taken from first-year examinations on the following courses at the Polytechnic of Central London:

B.A.S.S. = B.A. (Social Science), C.N.A.A. (full-time course)
B.A.S.S. (P/T) = B.A. (Social Science), C.N.A.A. (evening course)
B.A.B.S. = B.A. (Business Studies), C.N.A.A.
A.F.C. = Accountancy Foundation Course
H.N.D. = Higher National Diploma in Business Studies

I would also like to thank Cambridge University Press for permission to quote the normal distribution tables from the *Cambridge Elementary Statistical Tables*.

Glossary of Notation

The following list gives the notation in the order it appears in the book.

L_1	Lower limit of group
L_2	Upper limit of group
f	Frequency of items in group
Z	Mode
Q_1	Lower quartile
Q_3	Upper quartile
Q_2	Median (as middle quartile)
M	Median
D_1	Lowest decile
D_9	Highest decile
n	Number of items in sample
Σ	(sigma, upper case) Sum of
$\overline{x}$	Mean of sample data
C	Class interval
A	Assumed mean
x'	Value of x adjusted for assumed mean
σ	(sigma, lower case) Standard deviation of population data
X, Y	Horizontal and vertical co-ordinates respectively of values in regression and correlation problems; horizontal and vertical axes
$\overline{X}$, $\overline{Y}$	Mean values of X and Y
$\hat{Y}$	Expected value of Y
a	Intercept on Y axis of regression line
b	Regression coefficient — gradient of regression line
r	Correlation coefficient of sample data
d	Difference in ranks
r^2	Coefficient of determination
ρ	(rho) Correlation coefficient of population data
p_n	Current year price
p_o	Base year price
q_n	Current year quantity
q_o	Base year quantity
W	Weight
I	Price relative
T	Trend
S	Seasonal factor
R	Residual factor
P(A)	Probability that event A will occur
P($\overline{\text{A}}$)	Probability that event A will not occur

$P(A+B)$	Probability that event A *or* event B *or* both will occur
$P(AB)$	Probability that both events A and B will occur together
$P(B/A)$	Probability that event B will occur, given that event A has occurred
Φ	(phi, upper case) Area to the left of the ordinate of the normal curve
z	Standardised normal variable
$n!$	Factorial n, equal to $1 \times 2 \times 3 \ldots \times n$
p	Probability of success, or proportion of successes
q	Equal to (1 – p). Probability of failure, or proportion of failures
$\geqslant$	Greater than or equal to
$>$	Greater than
e	Mathematical constant equal to 2.718 3
m	Mean of Poisson variable when data are obtained from a sample
p	Sample estimate of proportion
π	(pi) Population value of proportion
μ	(mu) Mean of population data
s	Standard deviation of sample data
$\neq$	Not equal to
$<$	Less than
$\leqslant$	Less than or equal to
χ^2	Chi squared (see Chapter 11)
O	Observed frequency
E	Expected frequency
r	Number of rows in contingency table
c	Number of columns in contingency table

1 Introduction to statistics — presentation of statistics — error and approximation

The subject of statistics deals with the collection of data; the presentation of data; the analysis of data and the interpretation of the results of the analysis of data. Examination questions in statistics usually cover all these areas. There is a tendency for some students to concentrate on the analysis of data and neglect the other areas — this is unwise.

Collection of data

The first step in a statistical analysis is to obtain the data. *Primary data* are data collected by means of a census or survey. *Secondary data* are data obtained from published or unpublished results of someone else's survey or census. If you need data on a particular topic, it is always best to see if the data have already been collected, as this is very much cheaper than having to collect the data yourself.

Secondary data

A vast amount of data is collected from various sources — individuals, companies local authorities etc., and is published.

Government departments, trade associations, the United Nations, the European Economic Community (the Common Market) etc., collect and publish statistics. The following, all published by H.M.S.O., relate to the U.K.:

Monthly Digest of Statistics
The Annual Abstract of Statistics
Department of Employment Gazette
Trade and Industry
Economic Trends
Financial Statistics
Social Trends
Population Trends
National Income and Expenditure
Balance of Payments
General Household Survey
Family Expenditure Survey
Population Census Reports
Business Monitor

These contain statistics on numerous topics. College libraries and public libraries contain many of these publications, and it is very useful actually to look at some of these publications to see what data are available and how they are presented. If you wish to find out where to look for data on a particular topic, the H.M.S.O. publication *Guide to Official Statistics* is very comprehensive.

Sometimes the published statistics do not provide the information you require; quite often a telephone call to the government department concerned may help. The government collects a vast amount of statistics but publishes only some of the data collected. Government departments will often provide further information on request, sometimes without charge, sometimes at the cost of retrieval of the information from computer storage.

Primary data

If you cannot find any published statistics on a topic in which you are interested or cannot gain access to any unpublished statistics and you want to obtain data on the topic, you have no choice but to conduct a survey or employ a market research agency to do the survey for you. Statistics obtained from such a survey are called *primary statistics*. Survey methods are discussed in Chapter 12.

Presentation of statistics

When we have a mass of figures e.g. sales commission of 100 representatives, weekly wages of 1 000 employees, we can gain no immediate impression from this mass of data. It is necessary to present data in such a way that they can be comprehended by specialists and laymen alike. In many situations, diagrammatic presentation of numerical data is the most effective way of communicating statistical information to the public. Presentation of statistics is dealt with in more detail later in this chapter.

Analysis of data

The collected data usually need to be processed. This is covered by the general heading of *analysis of data*. Analysis of data involves procedures such as calculating means, correlation coefficients, index numbers and so on. The various calculations are dealt with separately in Chapters 2 to 11.

Interpretation of the results

The analysis of data involves the student in producing a numerical result. In many cases this process is a rote application of a technique, and with practice students usually become quite proficient at these techniques. Examiners are, however, only too well aware that students frequently obtain a correct numerical answer to a statistics question but have no real understanding of what the calculation involves and/or what the result means. Most examination questions on analysis of data involve two parts: (i) a calculation, (ii) an interpretation element; in some cases the proportion of the marks given for understanding the techniques and interpretation of the result can be more than half the marks for the question. It is, therefore, very important to be able to appreciate what you are attempting to do in the calculation and to make appropriate comments on the results. Chapters 2 to 11 attempt to make you proficient at doing the various techniques *and* to understand what the various techniques involve. Unfortunately, there is a tendency for some students to concentrate on the former and neglect the latter. It is probably true that you can pass if you are good at doing the numerical part; but you will not obtain good marks unless you can deal with the interpretation part of the question satisfactorily.

Presentation of data

If we are given a mass of data, this is not usually very meaningful. For example, consider the following data:

Monthly sales commission (£) of 100 representatives of a company

46	51	31	49	36	53	55	43	63	60
35	67	38	46	35	47	41	36	44	54
51	59	40	45	66	36	55	41	57	53
47	44	39	50	41	52	45	23	53	39
64	51	53	48	32	45	52	61	61	25
44	48	51	48	80	46	36	58	57	43
54	35	60	46	44	56	44	52	50	55
58	56	56	43	40	32	42	52	45	47
60	46	66	52	52	69	49	57	61	47
71	42	50	73	52	58	33	51	54	45

The usual procedure is to put the data into a table where there are clear categories to place each observation. In the table below the categories are: under £30, £30 and under £35, £35 and under £40, etc. £30–£35, £35–£40 etc., is *not* satisfactory because £35 could be placed in either category. The most convenient way of finding the total for each category is to use a tally, marking 1 for each entry, and to facilitate counting, it is useful to mark every fifth item entered by a diagonal stroke, i.e. 1111

Monthly sales commission (£) of 100 representatives

Range of commission (£)		*Number of representatives*
Under 30	11	2
30 and under 35	1111	4
35 and under 40	1111 1111	9
40 and under 45	1111 1111 1111	15
45 and under 50	1111 1111 1111 1111	19
50 and under 55	1111 1111 1111 1111 11	22
55 and under 60	1111 1111 111	13
60 and under 65	1111 111	8
65 and under 70	1111	5
70 and over	111	3
	Total	100

Source: *Company Records*

The table above is called a *frequency table.* (The tally marks are usually omitted in the published result.) When drawing up a table, the following points should be noted:

i The number of ranges to use depends on the data, but between 5 and 15 is usually acceptable.
ii Clearly define the units (in the table above, pounds).
iii Give the table a title.
iv Give the source of the data.
v Give column totals and, if the table involves a number of columns of data, when appropriate give row totals.
vi Give footnotes where necessary to assist interpretation of any part of the table.

Discrete and continuous data

The categories 30 and under 35, 35 and under 40 etc. in the table above involve a class interval of £5 (or more strictly £4.99$\frac{1}{2}$). The limits of the class intervals need to be defined. When data is continuous, e.g., data involving time, weights, heights, distance etc., the limits of the class intervals are usually straightforward. For example, on data involving waiting times, the categories 0 and under 5 minutes, 5 minutes and under 10 minutes and so on have class limits which are clearly defined. When we have discrete data, e.g. number of marks in an examination, number of employees in an office, the class intervals may be taken as 10–19, 20–29, 30–39. The class limits, however, need to be defined. The usual practice in drawing a histogram of discrete data (histograms are explained on page 7) is to regard 10–19 as falling in the interval 9$\frac{1}{2}$ - 19$\frac{1}{2}$; the class limits of the class interval 10–19 are 9$\frac{1}{2}$ and 19$\frac{1}{2}$.

Exercise 1.1

A company administers an aptitude test to 100 applicants for a job with the company. The following are the times taken to complete a simple task for each applicant, measured to the nearest second.

44	92	72	45	85	61	66	46	59	67
52	40	93	54	52	64	65	44	51	66
92	58	74	42	43	56	46	52	45	56
68	40	48	76	71	99	51	72	52	56
69	58	40	76	70	42	52	46	73	59
41	55	74	66	64	47	58	46	52	54
63	89	87	41	57	68	59	81	51	82
60	67	68	97	57	47	53	61	52	49
47	86	55	54	48	85	45	84	53	49
47	70	78	58	96	54	62	60	57	58

Draw up a frequency table for the above data using classes of 40–49, 50–59 etc.

Tabulation of Data

Sometimes we are given data in the following form:
U.K. *merchant vessels of* 500 *gross tons and over*

In 1966 there were 2 319 vessels and the total gross tonnage was 20 522 thousand tons. In 1976, whilst the number of vessels had reduced to 1 573, the total gross tonnage had increased by 45.4%. In 1966 (1976 figures are given in brackets) there were 191 (104) passenger vessels, gross tonnage 1 971 (661) thousand tons; 1 637 (780) tramps and cargo liners, gross tonnage 10 694 (5 407) thousand tons; 491 (423) tankers, gross tonnage 7 857 (15 742) thousand tons. In 1966 bulk carriers were included with tramps and cargo liners. In 1976 bulk carriers where shown separately.

Source: *Department of Industry*

For many people, understanding the data given above is improved if the data are presented in tabular form as shown below:

U.K. merchant vessels of 500 gross tons and over

Type of merchant vessel	1966 *Number of vessels*	1966 *Tonnage (1 000 tons)*	1976 *Number of vessels*	1976 *Tonnage (1 000 tons)*
Passenger	191	1 971	104	661
Tramp and cargo liners }	1 637	10 694	780	5 407
Bulk carriers }			266	8 029
Tankers	491	7 857	423	15 742
Total	2 319	20 522	1 573	29 839

Source: *Department of Industry*

In order to complete this table some calculations had to be made. The total tonnage in 1976 was stated to be 45.4% more than that of 1966:

$$20\ 522 + \frac{45.4}{100} \times 20\ 522 = 20\ 522 + 9\ 317 = 29\ 839 \text{ tons}$$

The figures for bulk carriers in 1976 were found by subtraction, e.g. number of bulk carriers 1 573 – 104 – 780 – 423 = 266 and for tonnage 29 839 – 661 – 5 407 – 15 742 = 8 029.

Sometimes the presentation and understanding of a tabulation can be improved if appropriate percentages are incorporated in the tabulation. (Examination questions on tabulation often include such phrases as 'calculate appropriate secondary statistics' and this usually means find appropriate percentages.) In the example above it might be desirable to show the changes between 1976 and 1966 in percentage terms and this has been done below.

U.K. merchant vessels of 500 gross tons and over

Type of merchant vessel	1966 *No. of vessels*	1966 *Tonnage (1 000 tons)*	1976 *No. of vessels*	1976 *Tonnage (1 000 tons)*	*Percentage change* *No. of vessels*	*Percentage change* *Tonnage*
Passenger	191	1 971	104	661	–45.5	–66.5
Tramp and cargo liners }	1 637	10 694	780	5 407	–36.1	+25.6
Bulk carriers }			266	8 029		
Tankers	491	7 857	423	15 742	–13.8	+100.4
Total	2 319	20 522	1 573	29 839	–32.2	+45.4

Source: *Department of industry*

N.B. percentage changes were found as follows:

Passenger vessels decrease 191 – 104 = 87

$$\text{Percentage decrease} = \frac{87}{191} \times 100 = 45.5\%$$

Examination questions sometimes ask you to comment or to write a brief report. The main feature that the tabulation shows is that whilst the number of vessels has fallen by 32.2%, the tonnage has increased by 45.4%. Additionally, the large increase in tonnage of tankers and the reduction in tonnage of passenger vessels could also be mentioned.

Exercise 1.2

In England in 1971 there were 16 070 000 dwellings; of these 8 360 000 were owner-occupied, 4 500 000 were rented from local authorities, 2 410 000 were rented from private owners, and the remainder were held under other tenures. In Scotland in 1971 there were 1 800 000 dwellings; of these 540 000 were owner-occupied, 940 000 were rented from local authorities, 200 000 were rented from private owners, and the remainder were held under other tenures.

Source: *Social Trends*, 1972

a Tabulate this data, calculate appropriate secondary statistics and include these secondary statistics in your tabulation.
b Comment on your results.

Diagrammatic representation of data

The use of the frequency table showing the sales commission data and the tabulation of the U.K. merchant vessels data is one method of presenting data. Many people, however, find the tabular presentation of data confusing. Often a pictorial representation of data is much more effective.

Diagrams, charts and graphs have a visual impact. People can frequently appreciate a diagram, whereas a mass of figures has no impression. If we wish to show changes, a diagram often gives a clear demonstration of the changes, whereas percentage changes shown in a tabulation may not even be noticed.

Important note

When drawing diagrams, charts or graphs, certain rules should be followed to ensure that the diagrammatic presentation is informative:

i The diagram should have a title.
ii The source of the data for the diagram should be shown.
iii Axes should be clearly labelled and scales shown.
iv Where appropriate, shading or colouring should be used.
v A key should be provided where appropriate, i.e. if shading or colouring is used.
vi Footnotes should be included where necessary, but too many footnotes destroy the effect.

There are some important additional points to note if you want to score high marks on this type of question:

vii The diagram should be neat and tidy.
viii Use a ruler and a sharp, pointed pencil.

An untidy diagram, where the student uses the book of logarithm tables as a ruler, and a biro which smudges, no labels, no source, and no title, means few marks. This sort of question is easy so don't waste marks.

The government publications *Social Trends* and *Economic Trends* contain a good selection of the various forms of diagrammatic representation and are well worth looking at.

Histogram

The data on sales commission on page 2 may be shown in the form of a *histogram*.

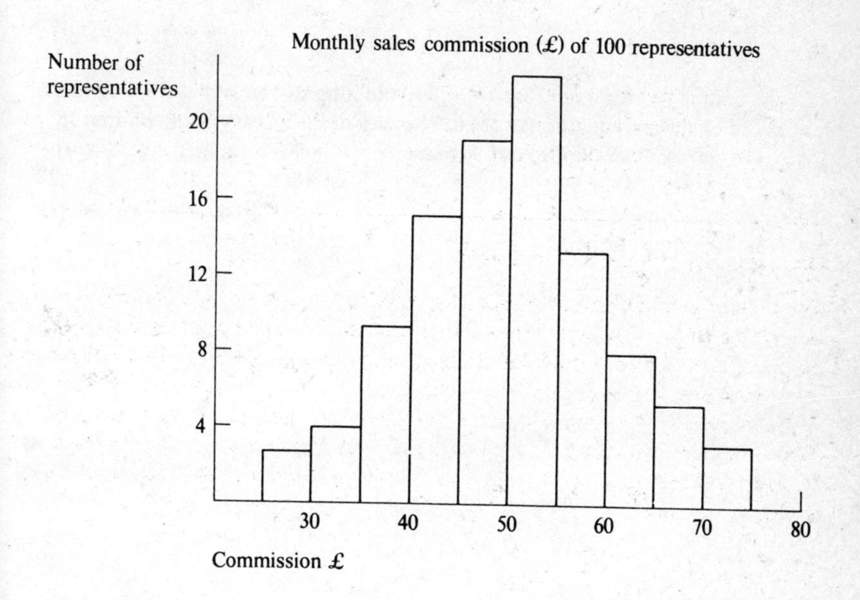

Figure 1.1 Source: *Company records*

The data are shown as a series of rectangles. The area of each rectangle is proportional to the number of representatives.

Exercise 1.3

Draw a histogram for the data in Exercise 1.1

In the example on sales commission the class intervals were all taken to be £5. Special care has to be taken when drawing a histogram when the class intervals are unequal. Consider the following example:

Age distribution of coloured persons

Age of person	*Percentage*
0–15	38
16–19	8
20–24	8
25–34	15
35–44	16
45–54	9
55–59	2
60 and over	4

Source: *General Household Survey*, 1975

If the data is plotted as in Fig. 1.2, the visual impression is that there is a large number of children under the age of 16. The area of the rectangle for children $38 \times 16 = 608$ is larger than the area of all the other rectangles combined e.g. $(8 \times 4) + (8 \times 5) + (10 \times 15) + (10 \times 16) + (10 \times 9) + (5 \times 2) + (10 \times 4) = 522$. However, the actual number of children under the age of 16 at 38% is, of course, less than the total population 16 and over. Clearly this presentation is unsatisfactory and inaccurate.

Age distribution of coloured persons

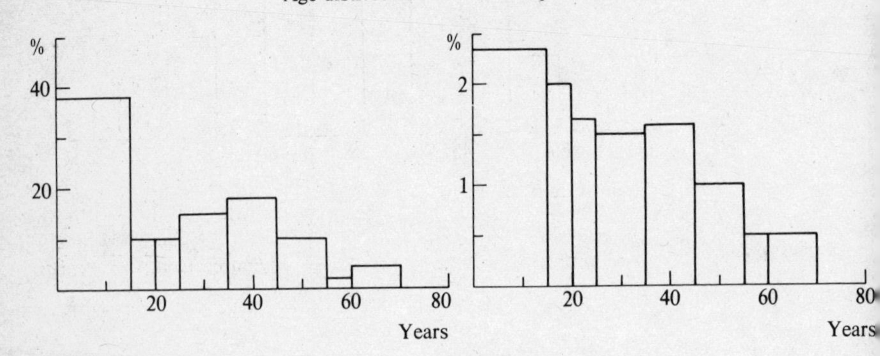

Figure 1.2

Source: *General Household Survey*

Figure 1.3

Source: *General Household Survey*

To obtain a better presentation, the standard procedure is to divide each percentage by the number of years i.e. for the first group $\frac{38}{16} = 2.375$

for the second group $\frac{8}{4} = 2$, for the third group $\frac{8}{5} = 1.6$

and so on. For the last group those aged over 60, there is no information about the spread of years. 10 years have been taken as a suitable figure but this is arbitrary.

Thus for the last group $\frac{4}{10} = 0.4$.

The result is plotted in Fig. 1.3. It will be noted that the figure for the under 16 age group of 2.375 assumes that for each age 0, 1, 2 . . . . . 15 there are 2.375% of the total.

Frequency polygon

If we take the data on sales commission and take the mid-points of the top of each rectangle and join these points, we obtain a *frequency polygon*. To draw a frequency polygon there is no need to construct a histogram, but if you do not draw a histogram you have to remember that you join the mid-points of the class intervals. A frequency polygon superimposed on a histogram is shown in Figure 1.4.

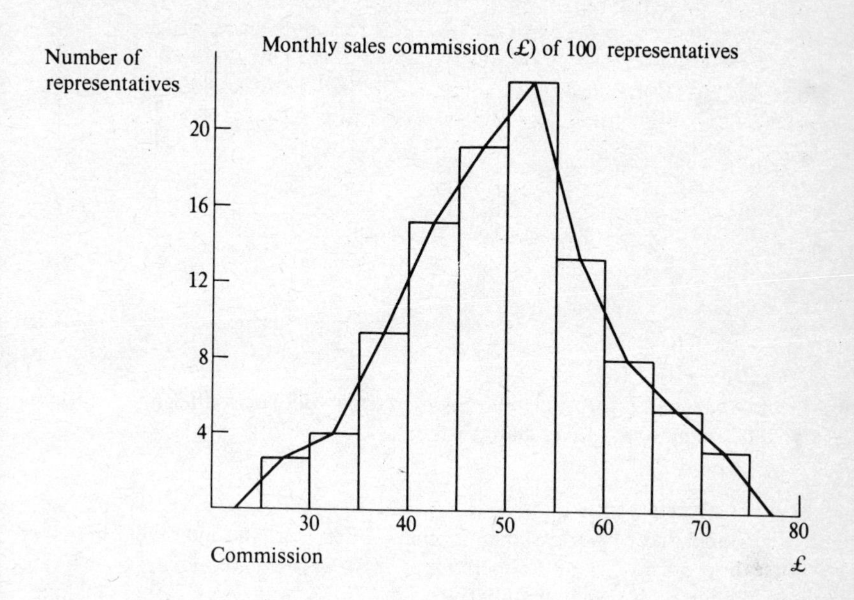

Figure 1.4 Source: *Company records*

Exercise 1.4

Draw a frequency polygon for the data of Exercise 1.1.

Bar charts

When data are in the form of a frequency table we use a histogram to present data diagrammatically. However, the data on merchant vessels on page 4 can be shown diagrammatically in various forms of bar chart.

Simple bar chart

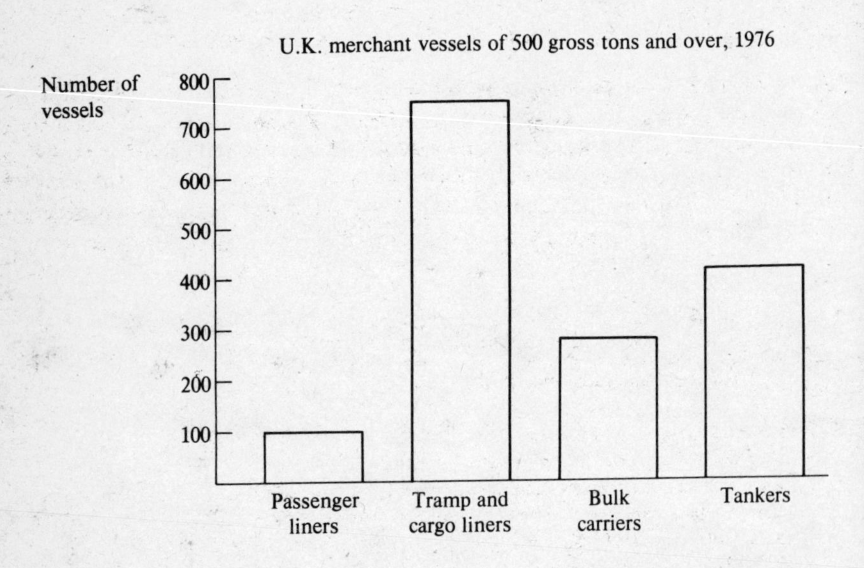

Figure 1.5 Source: *Department of Industry*

A simple bar chart is formed by rectangles of equal width whose heights are proportional to the number of merchant vessels.

Multiple bar charts
Suppose now that we wish to compare changes between 1966 and 1976. There is of course the problem that in 1966 there was no separation of bulk carriers from tramp and cargo liners. To make a comparison, for 1976 bulk carriers have been included with tramp and cargo liners. (See Figure 1.6)

Component bar charts
These show the break-down of the total into its components, in the case of merchant vessels, into passenger liners, tankers, and others. This presentation shows actual figures, so the bars for 1966 and 1976 are of different length. It is easy to make comparison of the totals but less easy to compare changes in the individual components. Component bar charts, as with all other bar charts, can be shown either vertically or horizontally. (See Figure 1.7)

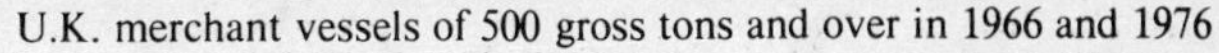

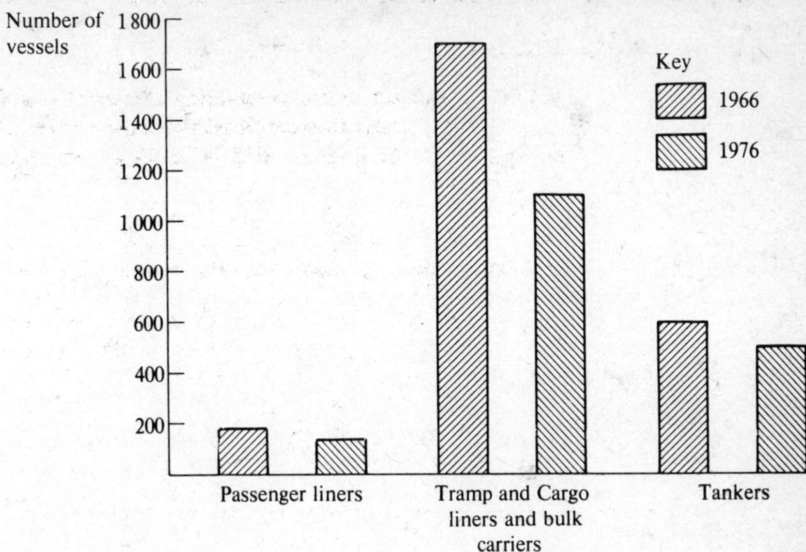

Figure 1.6 Source: *Department of Industry*

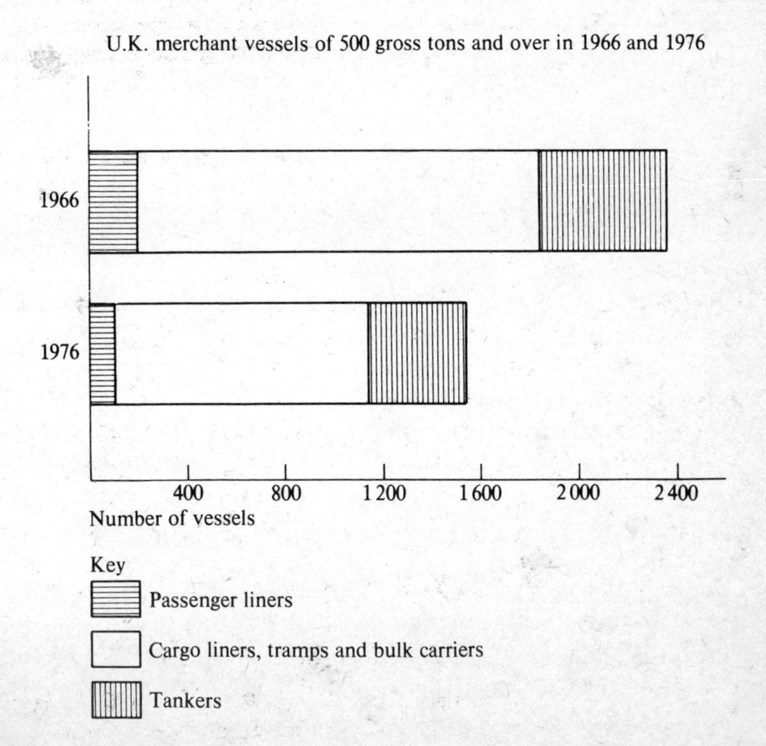

Figure 1.7 Source: *Department of Industry*

Percentage component bar charts

In comparing 1966 with 1976, the percentage component bar charts are drawn of equal length. The emphasis in the diagram is on the components and the change in the components in percentage terms. The first step is to change the data to percentages of the total.

	1966		1976	
Type of merchant vessel	*Number*	%	*Number*	%
Passenger liners	191	8.2	104	6.6
Tramps, cargo liners and bulk carriers	1 637	70.6	1 046	66.5
Tankers	491	21.2	423	26.9
Total	2 319	100.0	1 573	100.0

U.K. merchant vessels of 500 gross tons and over in 1966 and 1976

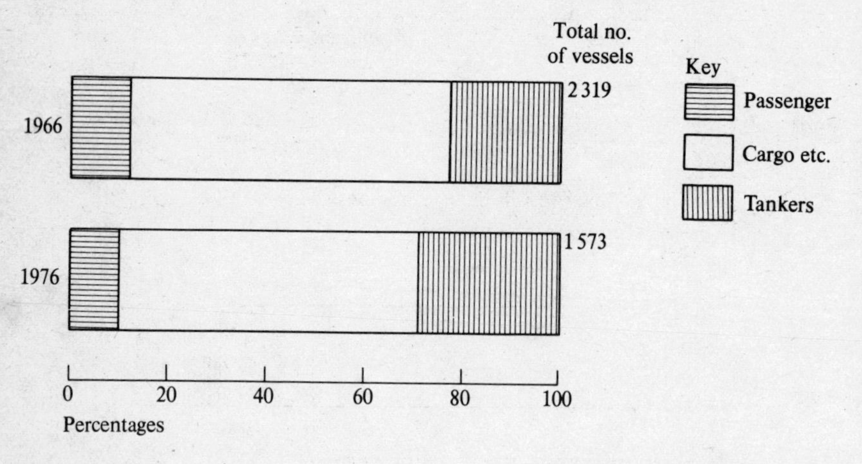

Figure 1.8 Source: *Department of Industry*

The percentage component bar chart does not allow comparisons of totals between the two years to be made, as the bars are of equal length. However, if the total for each year is placed at the end of the bar, such comparisons can be made.

Pie charts

An alternative method of showing the component parts is to use a pie chart. The various components are represented as sectors of a circle. The angles of the sectors are proportional to the size of the component. To find the angle for the sector corresponding to passenger vessels in 1966, we note that for a complete rotation, the angle is 360° and thus the angle for passenger vessels is $360 \times \frac{191}{2\ 319} =$ 29.7° and similarly for the other components.

Type of merchant vessel	1966		1976	
	Number	*Angle*	*Number*	*Angle*
Passenger liners	191	29.7	104	23.8
Tramps, cargo liners and bulk carriers	1 637	254.1	1 046	239.4
Tankers	491	76.2	423	96.8
Total	2 319	360.0	1 573	360.0

Source: *Department of Industry*

U.K. merchant vessels of 500 gross tons and over

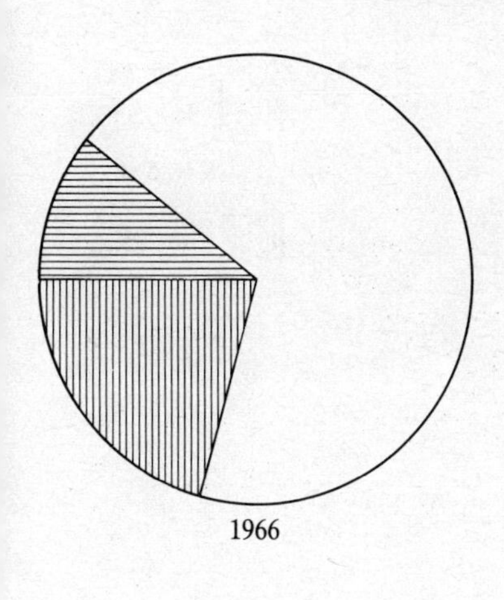

Figure 1.9

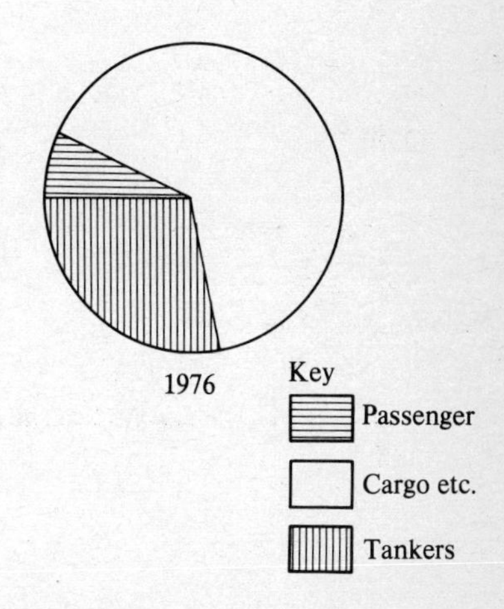

Figure 1.10

Source: *Department of Industry*

In drawing Figures 1.9 and 1.10 the radii of the two circles are in the ratio of the number of vessels in the two years, i.e. 2 319 to 1 573 or 1.474 to 1. Since the area of a circle is πr^2 the ratio of the area of Figure 1.9 to Figure 1.10 is $(1.474)^2$ to 1 or 2.173 to 1. As the eye looks at the *area* of the circle rather than the radius, this presentation is misleading. To ensure that the *areas* are in the ratio 1.474 to 1 we have to take the square root of 1.474 (1.214) and draw the circles with the radii in the ratio 1.214 to 1. This is done in Figures 1.11 and 1.12.

Pictograms

These are also called *ideograph*, *symbol* or *isotype charts*.

Symbols representing each variable are used, the symbol being easily recognised. For example we could compare the total number of vessels in 1966 and 1976 by drawing a number of ships. (See Figure 1.13)

U.K. merchant vessels of 500 gross tons and over

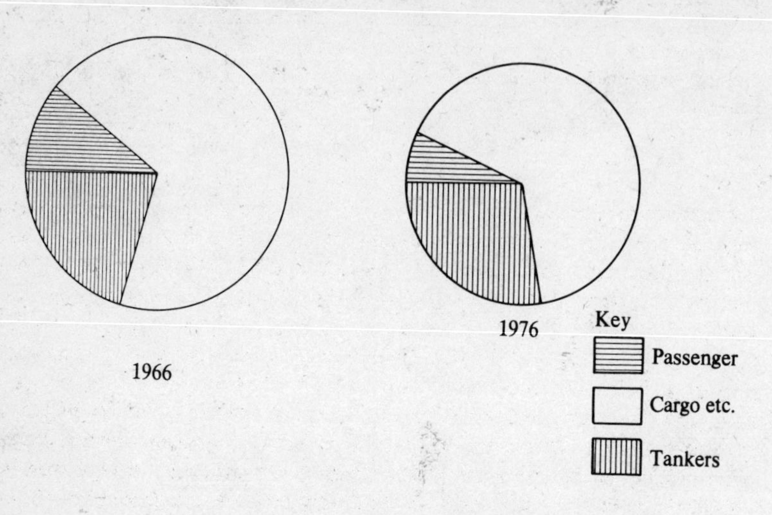

Figure 1.11

Figure 1.12

Source: *Department of Industry*

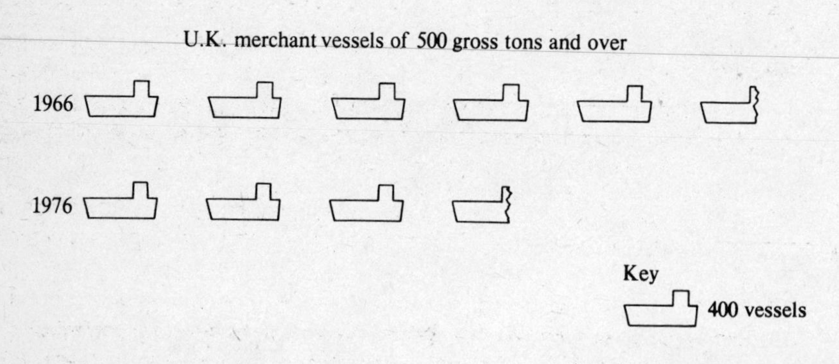

Figure 1.13

Source: *Department of Industry*

This form of presentation is very popular. A small defect is that fractions of the pictogram are not easy to interpret. There is also a danger of misrepresentation. For example, instead of Figure 1.13, 1966 could be represented by a ship 2.319 units long (1 unit for each 1 000 vessels) and 1976 by a ship 1.573 units long.

The visual impression is that the reduction from 1966 to 1976 is much greater than that indicated in Figure 1.13. If the ships are drawn three-dimensionally, the visual impact is greater. Occasionally, advertisers and even the U.K. Government (e.g. in *Britain and Europe*, page 10, July 1971) have used the form of presentation illustrated by Figure 1.14 but it is bad practice.

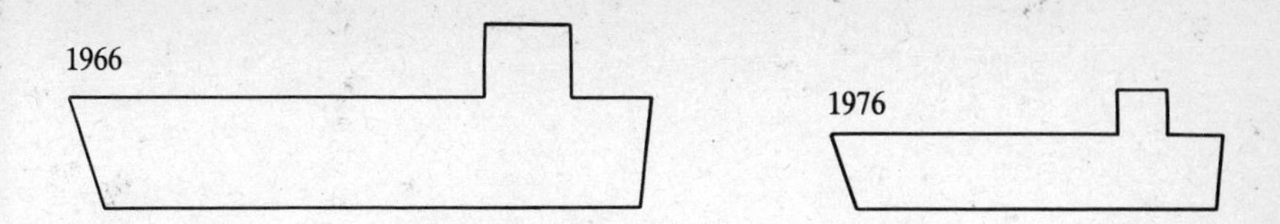

Figure 1.14

Examination papers sometimes ask questions such as: 'Use a suitable diagram to illustrate the following data'. The diagram to use depends upon the data and the aspects of the data which need to be illustrated. The choice is usually between a component bar chart, a percentage bar chart or a pie chart. You should go into the examination room with ruler, compasses, protractor, coloured pens or pencils and a rubber. If you have no protractor or compasses, a pie chart should be avoided. If you consider that the important comparison is between the totals for, say, two different sets of data and that the relative proportion of the components is less important, then a component bar chart is suitable — you can always enter the percentage for each component in the chart. If the important comparison is the relative change in the proportions of the components, then a percentage bar chart or a pie chart can be used. If there are a large number of small components, a percentage bar chart is better than a pie chart.

Exercise 1.5

Draw a (a) simple bar chart, (b) multiple bar chart, (c) component bar chart, (d) percentage component bar chart, (e) pie chart, (f) pictogram to illustrate the data of Exercise 1.2.

Graphs

Some data come in the form of a *time series*, where we have values for a number of consecutive years. The following data on colour television licences is an example:

Year	*Number of colour television licences (thousands)*
1968	75
1969	197
1970	528
1971	1 305
1972	2 815
1973	5 007
1974	6 824
1975	8 294
1976	9 569

Source: *Post Office*

When we have time series data, time is always plotted on the horizontal axis. The above data are shown in Figure 1.15.

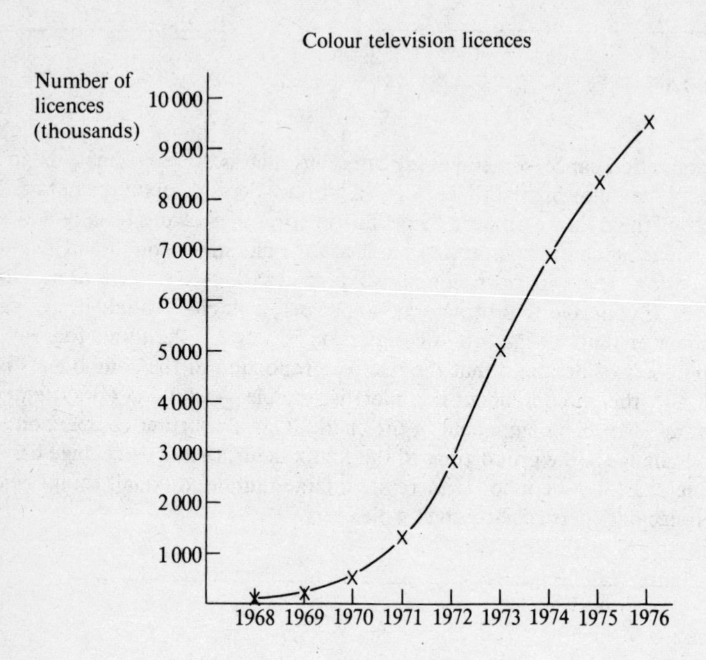

Figure 1.15 Source: *Post Office*

Figure 1.15 is called an *arithmetic scale graph*; equal vertical amounts show equal changes in the variables. Sometimes it is advisable to consider *rate* of change rather than absolute change; to do this we use *semi-logarithmic graph paper*. With this graph paper equal vertical amounts show equal percentage or relative changes. Figure 1.16 shows the television licence data on semi-logarithmic graph paper. As this graph is decreasing, this shows that the rate of increase of television licences is falling, whereas Figure 1.15 shows that the actual numbers are increasing.

In the examination room you may or may not be provided with semi-logarithmic graph paper. If you are provided with the paper, the vertical axis is usually marked 1 2 3 . . . 9 for each cycle. For the data on colour television licences, as the first reading is 75, it is necessary to place a zero after each number in the first cycle, i.e. 10, 20, 30, . . . 90 and for the next cycle two zero's i.e. 100, 200, 300 . . . 900 and so on. It will be noted that semi-logarithmic graph paper has no zero value, and negative values cannot be plotted.

If you are not provided with semi-logarithmic graph paper, it is necessary to take logarithms of the data.

Year	1968	1969	1970	1971	1972
Number	75	197	528	1 305	2 815
Logarithm	1.8751	2.2945	2.7226	3.1155	3.4495

Year	1973	1974	1975	1976
Number	5 007	6 824	8 294	9 569
Logarithm	3.6996	3.8341	3.9188	3.9809

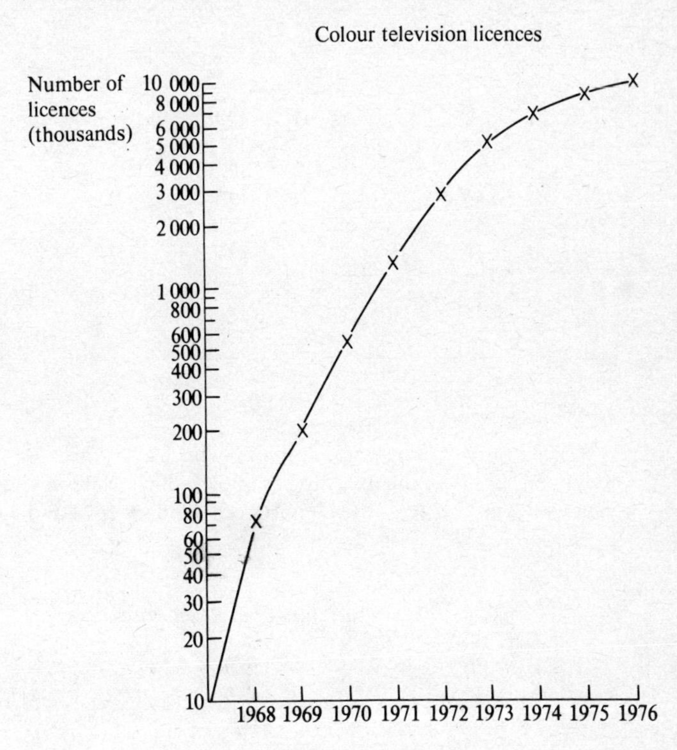

Figure 1.16 Source: *Post Office*

We now plot on ordinary graph paper. Against 1968 we plot 1.88 (rounding 1.8751 up to two decimal places), for 1969 we plot 2.29 and so on. The shape of the curve should be the same as in Figure 1.16.

Examination questions on this topic often ask you to 'plot using appropriate graph paper'. The choice is effectively between ordinary graph paper or semi-logarithmic graph paper. In deciding which graph paper to use it is important to consider what aspects of the data need illustrating. For example, if you were given data on turnover and profit of a company for a number of years, the important comparison may be to see whether the rate of increase of profits is comparable to the rate of increase of turnover. In this situation, semi-logarithmic graph paper should be used. On the other hand if, during one year, the company had zero profits or made a loss, semi-logarithmic graph cannot be used as zero and negative values cannot be plotted.

Exercise 1.6

a Discuss the relative advantages and disadvantages of (i) ordinary graph paper, (ii) logarithmic graph paper.

b Use a suitable graph to display the following data and comment on your results.

Year	*Sales* (£m)	*Operating profit* (£m)	*Year*	*Sales* (£m)	*Operating profit* (£m)
1967	43	4	1972	83	9
1968	47	4	1973	99	13
1969	59	5	1974	121	15
1970	62	6	1975	162	19
1971	72	7	1976	211	22

Source: *Company Accounts* (R.S.A., 1978)

Misleading graphs

By unsuitable choice of origin and scale a graph may be drawn which may give a misleading impression. Consider the following example, which relates to the movements of the General Index of Retail Prices between January 1974 and January 1978.

Date	*General Index of Retail Prices*
January 1974	100
January 1975	119.9
January 1976	147.9
January 1977	172.4
January 1978	189.5

Source: *Department of Employment Gazette*

Figures 1.17 and 1.18 plot the same data but the rise in prices looks much more rapid in Figure 1.17. If you are arguing that the rise in prices has been rapid you would choose Figure 1.17. If you were trying to give the impression that the rise in prices was not all that bad, you could use Figure 1.18.

In Figure 1.17 the effect of a rapid rise in the graph is obtained by omitting the range 0 – 100, thereby expanding the vertical scale, and by compressing the scale on the horizontal axis, compared with Figure 1.18.

Lorenz curve

The Lorenz curve is used to give a visual impression of the degree of inequality. It is frequently used to illustrate the inequality of the distribution of income or wealth,

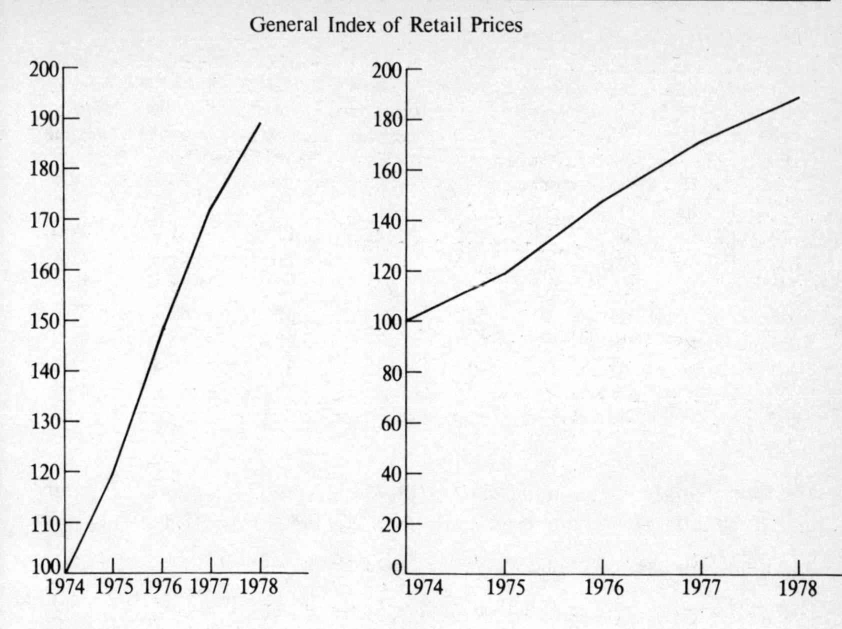

Figure 1.17

Figure 1.18

Source: *Department of Employment Gazette*

i.e. that wealth is concentrated in the hands of a relatively small number of individuals. The Lorenz curve can also be used to show the degree of concentration within an industry, i.e. to show that the manufacture of certain products is concentrated in a small number of large firms and that a large number of small firms make a relatively low proportion of the total output. To illustrate the method of constructing a Lorenz curve, consider the following data:

Distribution of personal incomes before tax, 1974–75

£	*Number of tax units* (100 *thousand*)	*Income before tax* (£100m)
under 1 000	66	48
1 000 and under 1 500	47	53
1 500 and under 2 000	35	52
2 000 and under 2 500	35	65
2 500 and under 3 000	31	69
3 000 and under 4 000	40	111
4 000 and under 5 000	17	58
5 000 and under 10 000	12	56
10 000 and over	2	15
Total	285	527

Source: *Central Statistical Office*

	1	2	3	4	5	6
Range of income £	*No. of tax units*	*Income*	*Cumulative no. of tax units*	*Cumulative income*	*% Cumulative numbers*	*% Cumulative income*
under 1 000	66	48	66	48	23.2	9.1
1 000 and under 1 500	47	53	113	101	39.6	19.2
1 500 and under 2 000	35	52	148	153	51.9	29.0
2 000 and under 2 500	35	65	183	218	64.2	41.4
2 500 and under 3 000	31	69	214	287	75.1	54.5
3 000 and under 4 000	40	111	254	398	89.1	75.5
4 000 and under 5 000	17	58	271	456	95.1	86.5
5 000 and under 10 000	12	56	283	512	99.3	97.2
10 000 and over	2	15	285	527	100.0	100.0
	285	527				

In the table above the actual data is placed in columns 1 and 2. Column 3 cumulates the data of column 1, i.e. 113 = 66 + 47, 148 = 66 + 47 + 35 and so on. Column 5 expresses column 3 as a percentage of the total 285: $\frac{66}{285} \times 100 =$ 23.2; $\frac{111}{285} \times 100 = 39.6$

Columns 4 and 6 are calculated in the same way. Columns 5 and 6 show that 23.2% of tax units (the number of tax units and the number of people are nearly the same) earn 9.1% of the total income. If we had perfect equality of the distribution of income, 23.2% of the people would earn 23.2% of the total income.

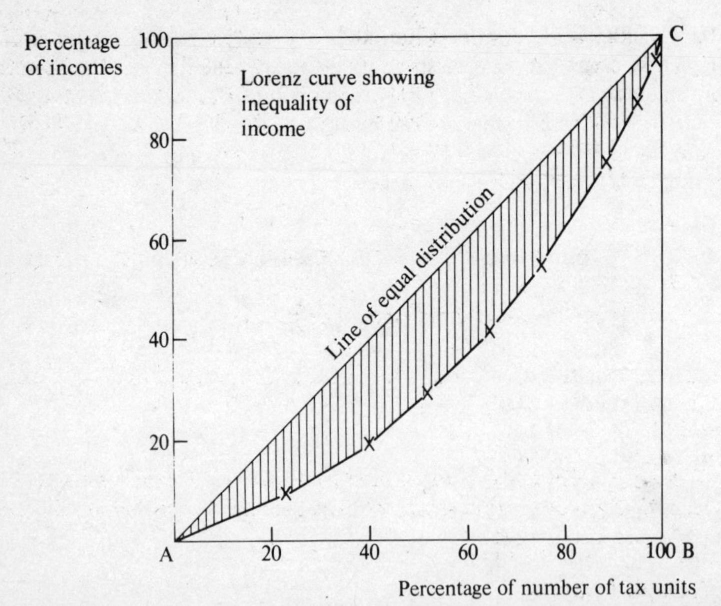

Figure 1.19 Source: *Central Statistical Office*

Columns 5 and 6 are plotted: cumulative numbers along the horizontal axis; cumulative income along the vertical axis, as in Figure 1.19.

The line drawn diagonally is the line of equality and would be the line of the graph if we had perfectly equal distribution of income. If we had perfect inequality of income, the line would be along the line AB-BC. Figure 1.19 shows that there is a degree of inequality of income. A measure of the degree of inequality is given by the shaded area in relation to the area of the triangle ABC. This measure is called the *Gini coefficient.*

$$\text{Gini coefficient} = \frac{\text{Shaded area}}{\text{Area of triangle ABC}}$$

When we have perfect equality, the Gini coefficient is zero; when we have perfect inequality, the Gini coefficient is 1. For the data above, we need to find the shaded area — this can be done by counting the number of small squares within the shaded area. The Gini coefficient is approximately 0.315.

Exercise 1.7

Distribution of income before tax 1973/4

Range of incomes *Not under* (£)	*Under* (£)	*Number of tax units* (*thousand*)	*Income before tax* (*£ million*)
	595	3 579	1 800
595	1 000	5 499	4 513
1 000	1 500	4 488	5 561
1 500	2 000	4 220	7 360
2 000	2 500	3 800	8 509
2 500	3 000	2 764	7 534
3 000	5 000	3 114	11 311
5 000	10 000	523	3 448
10 000 and over		136	2 183
	Total	28 123	52 219

Source: *National Income and Expenditure*

From the above data draw a Lorenz curve.

Z chart

The Z chart (or Zee chart) is often used to display sales data. The sales data is formed into three component curves, which are roughly in the shape of the letter Z. The component curves are:

i The actual data
ii The cumulative data
iii The moving annual total

To illustrate the method of construction, consider the following data:

Sales of ABC Company (£1 000)

Month	*1977*	*1978*
January	32	37
February	29	31
March	38	45
April	49	58
May	57	62
June	75	79
July	88	102
August	79	88
September	49	61
October	43	57
November	39	50
December	46	53

i The actual data are the monthly sales for 1978.
ii We cumulate the data for 1978; this is shown in Column 2 below.
iii We find the moving annual total. The annual total for January 1978 is the total from February 1977 to January 1978 inclusive, i.e. 29 + 38 + 49 + ... + 46 + 37 = 629. The annual total for February 1978 is the total from March 1977 to February 1978, i.e. 38 + 49 + 57 + . . . + 37 + 31 = 631. An easier way to find the figure for February 1978 is to note that 631 = 629 + 31 – 29. If you proceed in this way you should obtain a figure of 723 for December 1978. This figure *must* be the same as the cumulative total; if it is not, an error has been made and you should check to find the error.

Month	(1) *Monthly sales*	(2) *Cumulative total*	(3) *Moving annual total*
January	37	37	629
February	31	68	631
March	45	113	638
April	58	171	647
May	62	233	652
June	79	312	656
July	102	414	670
August	88	502	679
September	61	563	691
October	57	620	705
November	50	670	716
December	53	723	723

The data are plotted in Figure 1.20.

The data show any seasonal factors, in this case a pronounced increase in sales in the summer months. The cumulative total shows the total sales for the year up to the months concerned, i.e. the figure of 502 for August shows that sales since January are £502,000. The moving annual total, as it involves a total for a year, has

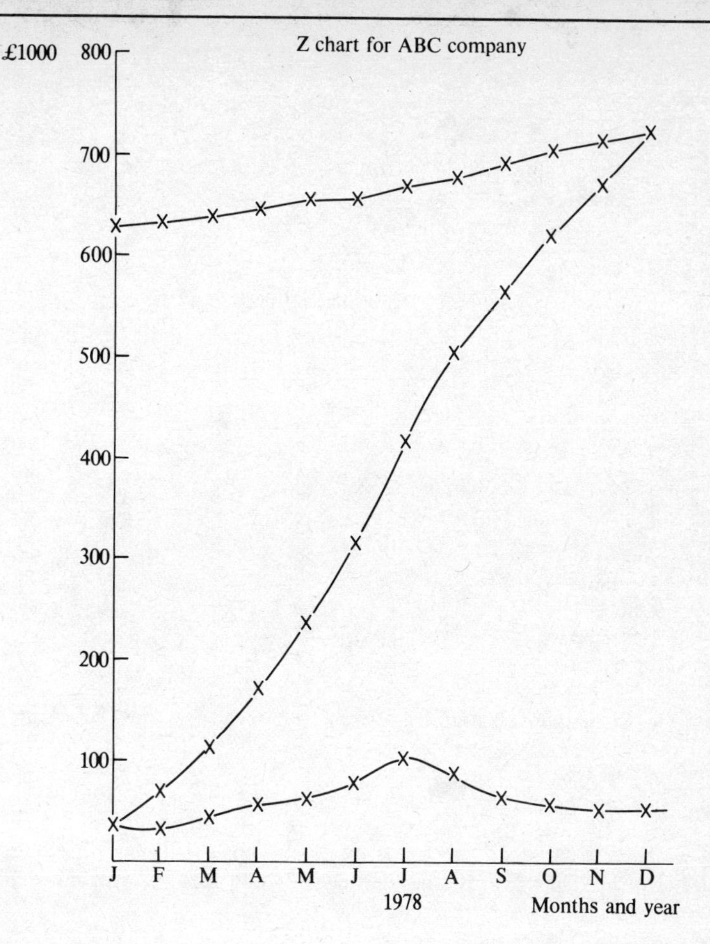

Figure 1.20 Source: *Company accounts*

eliminated the seasonal effects and thus shows whether sales are rising or falling. In this case sales are rising.

Accuracy and approximation

So far in this chapter we have represented in diagrammatic form data of various types, but how accurate are the data we have been given? If data are obtained from a sample, there is always the chance that the sample is not representative of the population and thus the sample data may not reflect the population data. Errors as a result of sampling are called *sampling errors*. How to deal with sampling errors is discussed in Chapter 10.

If data are from a census, the data are more accurate than from a sample, but there is not usually complete accuracy. In the case of the population census a number of people are accidently missed on census night. The statistics of the national income of the United Kingdom are not perfectly accurate; some components are more accurate than others. The government statisticians classify

Exercise 1.8

a Briefly discuss the usefulness of the Z chart (zee chart) to the businessman.

b Draw a Z chart of the following data:

Sales of XYZ Company (£1 000)

	1976	*1977*
January	68	60
February	76	62
March	84	69
April	97	84
May	64	58
June	47	40
July	32	28
August	56	45
September	57	48
October	60	46
November	62	50
December	64	52

c Comment on your Z chart. (R.S.A. 1978)

the accuracy of the various components by the gradings A, B and C: A means margin of error less than 3%, B error between 3% and 10%, C error more than 10%.

There is an additional accuracy error which often appears in students' answers to examination questions, and probably occurs in some of the answers to the worked examples and exercises in this book! This is *spurious accuracy*. There is a tendency to read off all the digits from the calculator display and give these as the answer; we then have results such as average income = 49.674 829 (students have a bad habit of omitting the £ sign). Often in calculating a mean some arbitary assumptions have to be made on the central values (these assumptions are discussed in Chapter 3) and giving the answer to the nearest penny, i.e. £49.67, is probably more accurate than the data and the analysis warrants. It is essential to avoid spurious accuracy when giving your results. The level of accuracy of the original data give an indication of the level of accuracy needed for the answer.

Errors of approximation

In some questions, e.g. Exercise 1.2, you are asked to calculate secondary statistics and in this exercise this means finding percentages. In this exercise the data has been given 'rounded' to the nearest 10 000. Thus the figure of 1 800 000 dwellings for Scotland could be between 1 795 000 and 1 805 000 (the last figure could have been given more accurately as 1 804 999). Likewise the figure of 940 000 dwellings rented from local authorities could be between 935 000 and 945 000. The figure given in the solution to Exercise 1.2 is $\frac{940\ 000}{1\ 805\ 000} \times 100 = 52.22\%$. However if we take the extreme values 1 805 000 for the number of dwellings and 935 000 for the number of dwellings rented from local authorities, the percentage is $\frac{935\ 000}{1\ 805\ 000} \times 100 = 51.80\%$. Similarly if we take 1 795 000 and 945 000, the percentage is 52.65%. Thus 52.22% gives the usually accepted value, but it has to be recognised that with the accuracy of the data given, the percentage could lie between 51.80 and 52.65. It should be noted that unless you are specifically asked to find the errors of approximation 52.22% (or perhaps more realistically 52%) is the answer to give.

The method of finding errors of approximation depends upon whether you have to multiply or divide. To illustrate the method consider the following example.

A firm employs on average 600 employees but the actual numbers employed fluctuate by 10 on either side of this figure. The average weekly pay to the nearest pound is £72. The expected number of days the factory is producing in a year is 240, but this can vary by 2 days either way. Find: (a) The total wage bill for the year; (b) The average wage bill for each day the factory is operating. In both cases give the limits of error.

a Total wage bill is 600 × 72 × 52 = £2 246 400
The highest total wage bill is 610 × 72.49 × 52 = £2 299 382.80
The lowest total wage bill is 590 × 71.50 × 52 = £2 193 620
(It should be noted that when we multiply we take the two highest figures for number employed and wages to obtain the highest total; we take the two lowest figures for number employed and wages to obtain the lowest total.)

b The average daily wage bill is total wage bill divided by the number of days = $\frac{£2\ 246\ 000}{240} = £9\ 360$

The highest average daily wage bill is the *highest* total wage bill divided by the *lowest* number of days = $\frac{2\ 299\ 382.80}{238} = £9\ 661$

The lowest average daily wage bill is the *lowest* total wage bill divided by the *highest* number of days = $\frac{2\ 193\ 620}{242} = £9\ 065$

Absolute error, relative error, percentage error

Absolute error is the difference between the central value and the limiting values. In the example above the absolute error is £2 246 400 – £2 193 620 = £52 780 on the total wage bill. It should be noted that the difference between £2 299 382.80 and £2 246 400 at £52 982.80 is different from the first value; this is because of the method of calculation and it is usual for the two absolute errors to be slightly different. Examiners usually accept either of these results.

Relative error is the ratio of the absolute error to the central value i.e.

$$\frac{52\ 780}{2\ 246\ 400} = 0.023\ 5$$

Percentage error is the relative error expressed as a percentage: 2.35%

Exercise 1.9

A Manufacturer, in making an article, has the following expenses:

Materials	£2,000 ± 5%
Wages	£4 000 ± £300

He plans to sell the articles made as follows:

No. of articles	1 000 ± 70
Price	£8 ± 6.25%

Find **a** Maximum and minimum expenses;
b Maximum and minimum receipts;
c Maximum and minimum profits;
d Maximum and minimum profit per article.

(I.O.S. 1972)

Revision Exercises

1.10 The analysis of a firm's order book gave the following value of orders received, values being recorded to the nearest £.

Orders

47	80	43	52	45	54	52	69
56	82	55	48	56	59	58	68
64	81	60	58	62	62	62	72
67	83	66	61	66	69	69	60
72	86	70	67	71	74	73	63
77	89	75	72	76	79	78	68

a Tabulate the data into the form of a frequency table using intervals of £5.
b State clearly the limits of your second interval.
c Draw a histogram of the data.
d Draw a frequency polygon of the data.

1.11 **a** Briefly discuss the relative advantages and disadvantages of the various methods of presenting data diagrammatically.
b Present the following data using a suitable method

U.K. central government receipts 1974

	£ *million*
Rent, dividends and interest	2 088
Taxes on income	12 141
Taxes on expenditure	8 360
National Insurance contributions	4 986

Source: *National Income and Expenditure*, 1975

(R.S.A. 1977)

1.12 For the following three sets of data use three *different* methods of presenting the data diagrammatically:

a Number of Current Colour Television Licences (1 000s)

Year	*Number*
1968	75
1969	197
1970	528
1971	1 305
1972	2 815
1973	5 007
1974	6 824

Source: *Post Office*

b University students: percentages attending various universities

	1953–4	*1973–4*
Oxford and Cambridge	18	9
London	22	14
Other	60	77

Source: *Department of Education and Science*

c Tenure of dwellings in 1973

	Percentage
Owner-occupied	52
Rented from local authorities	31
Rented from private owners	13
Other tenures	4

Source: *Department of Environment* (R.S.A. 1976)

1.13 Age of Indian immigrants in households in G.B.

Age in years	*Percentages*
0–4	13
5–9	8
10–14	8
15–19	8
20–24	9
25–44	35
45–59	12
60–64	3
65 or more	4
All	100

Source: *Census of Population 1966*

Represent the distribution by a histogram and comment on any unusual features.

1.14 **a** What are the advantages of semi-logarithmic graph paper over natural scale graph paper?

Illustrate your answer with examples.

b Plot the following data, relating to a firm's sales of cider, on an appropriate graph and comment on what your graph shows.

Sales of cider

Year	$\frac{1}{2}$ *gall jars (thousand)*	*pint bottles (hundred thousand)*	$\frac{1}{2}$ *pint cans (thousand)*
1964	300	64	2.0
1965	350	76	2.1
1966	420	90	2.2
1967	500	105	2.3
1968	780	145	2.9
1969	1 300	200	4.1
1970	1 450	250	6.8
1971	1 650	300	11.0
1972	1 750	340	14.0
1073	1 900	380	16.5

(B.A.B.S. 1974)

1.15 The following data were obtained from the 1970 Census of Production:

Manufacture of electrical applicances primarily for domestic use

Number of employees	*Number of establishments*	*Total sales* (£m)
1–99	221	16
100–299	30	24
300–749	22	53
750–999	6	31
1 000–1 499	3	16
1 500–2 499	7	59
2 500 and over	6	107

(R.S.A. 1974)

Draw a Lorenz curve and comment on your results.

1.16 **a** Describe the uses of the Z chart.

b Draw a Z chart for the following data:

Sales of ABC Company (£ thousands)

	1971	*1972*
January	39	46
February	33	44
March	41	52
April	49	57
May	56	62
June	63	70
July	71	83
August	53	57
September	42	46
October	37	39
November	36	35
December	53	60

c Comment on the chart.

1.17 A solid rectangular block has its measurements taken to the nearest 0.1 cm. It has a square base, whose edge is given as 3.5 cm, and its height is given as 5.0 cm. What does this information imply concerning the limits of
i the perimeter of the base,
ii the volume of the block?
What do you understand by the terms 'biased error' and 'unbiased error'?

2 Mode, Median, Quartiles, Deciles, Quartile Deviation, Range

Measures of average

When we have data we often wish to find the value of the 'average'. The most common form of 'average' and the most well known is the *arithmetic mean*. This is obtained by totalling the items and then dividing by the number of items. The arithmetic mean is discussed in detail in Chapter 3. Another form of average is the *median*. This is obtained by arranging the data in order of magnitude and taking the middle value. The median is discussed later in this chapter. A third measure, and the least used of the these measures of average, is the *mode*. The mode is that item which occurs most frequently. If you remember the phrase 'à la mode' — the most fashionable — this helps to remind you that the mode refers to the most frequently occurring item.

The Mode

As stated above, the mode is the item which occurs most frequently. An examination of the monthly sales commission of 100 representatives of a company shown on page 3 reveals that the commission that occurs most frequently is £52 and this is, therefore the value of the mode for this data.

The calculation of the mode is different if the data is given in the form of a frequency table. The sales commission data of page 3 can be given in the form of a frequency table.

Worked Example No. 2.1

Sales Commission (£)	*No. of representatives*
under 30	2
30 and under 35	4
35 and under 40	9
40 and under 45	15
45 and under 50	19
50 and under 55	22
55 and under 60	13
60 and under 65	8
65 and under 70	5
70 and over	3

There are two methods that can be used.

Graphical method

The first step is to draw a histogram of the data. Take the group with the largest frequency — this is called the modal group — and join lines from the top of the rectangle of the modal group to the top of the adjacent rectangles, as in Figure 2.1. A vertical line is then drawn downwards from the point of intersection of these two lines. Where this line meets the horizontal axis indicates the position of the mode. From Figure 2.1 it is evident that the mode is approximately £51.

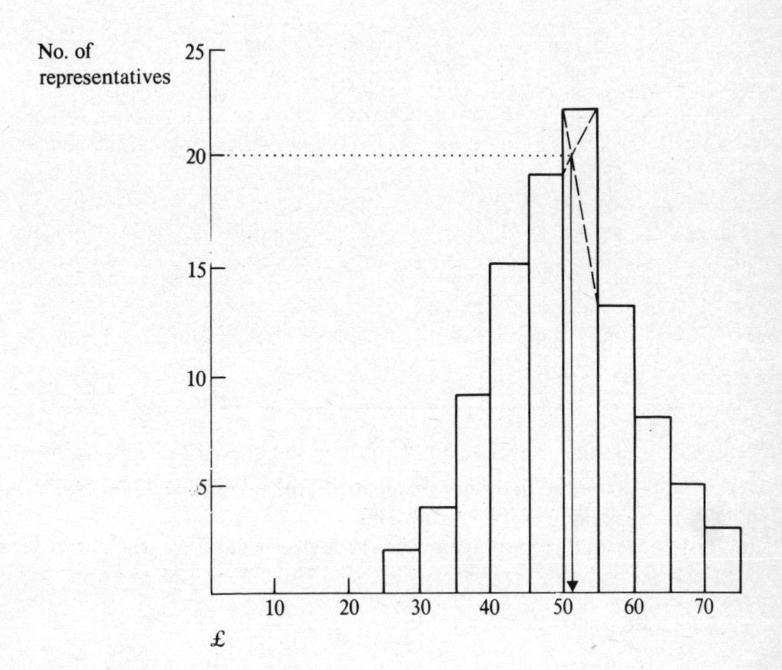

Figure 2.1

Calculation method

This is based on applying simple proportion to the graphical method of solution. It may be shown that the following formula gives the position of the mode.

$$L_l + \left(\frac{f_1 - f_0}{2f_1 - f_0 - f_2}\right)(L_2 - L_1)$$

Where L_1 is the lower limit of the modal group
L_2 is the lower limit of the group after the modal group
f_1 is the frequency of the modal group
f_0 is the frequency of the group before the modal group
f_2 is the frequency of the group after the modal group

In the example on sales commission, $L_1 = 50$; $L_2 = 55$; $f_1 = 22$; $f_0 = 19$; $f_2 = 13$. Substituting in the formula for the mode

$Z = \text{Mode} = 50 + \left(\frac{22 - 19}{2 \times 22 - 19 - 13}\right)(55 - 50) = 50 + \frac{3}{12} \times 5 = 50 + 1.25 = £51.25$

The calculation method and the graphical method give similar results.

Exercise 2.1

Distribution of examination marks:

Range of marks	*Percentage of candidates*
0 - 9	3
10 - 19	1
20 - 29	6
30 - 39	13
40 - 49	25
50 - 59	27
60 - 69	14
70 - 79	5
80 - 89	6

Find the mode.

Median

If a set of data is arranged in order of magnitude, the middle value, which divides the set into two equal groups, is the median.

Suppose that we take a random sample of 15 companies and note their profits for the last financial year.

Worked Example No. 2.2

Profits of 15 companies (in thousands of pounds)

27, 6, 38, 102, 54, 39, 70, 33, 73, 152, 1604, 380, 134, 46, 17

We arrange the data in order of magnitude:

6, 17, 27, 33, 38, 39, 46, 54, 70, 73, 102, 134, 152, 380, 1604

The middle value — the eighth item — is 54. The median profit for this sample of companies is £54 000.

A more common average is the arithmetic mean (discussed in chapter 3) which is familiar to all. The mean profit of these companies is

$$\frac{6 + 17 + 27 + 33 + 38 + 39 + 46 + 54 + 70 + 73 + 102 + 134 + 152 + 380 + 1\,604}{15}$$

$$= \frac{2\,775}{15} = 185 \text{ (£185 000)}$$

Thus the average profit, as measured by the arithmetic mean, is £185 000, a figure considerably higher than the median. This is due to the fact that two companies had relatively large profits compared with the other companies. It may be argued, therefore, that the median company profit at £54,000 is more representative of the companies of the sample. In income distributions it is more usual to give the median income than the mean income. Union negotiators frequently use the median income as the typical income of their members when negotiating with employers.

Exercise 2.2

Weekly earnings of a random sample of 15 employees of a company:
62, 42, 73, 80, 182, 78, 69, 103, 92, 84, 130, 58, 170, 71, 97
Find the median earnings.

Quartiles

The median together with the two quartiles divide the distribution into four quarters. The median is the value such that half the observations are greater than the median and half the observations are less than the median. The lower quartile, designated Q_1, is the value such that three quarters of the observations are greater than Q_1 and one quarter of the observations are less than Q_1. In our example on the profits of companies Q_1 would be the fourth item — 33. The upper quartile, designated Q_3, is the value such that one quarter of the observations are greater than Q_3 and three quarters of the observations are less than Q_3. In our example on the profits of companies, Q_3 would be the twelth item — 134.

The quartiles measure the spread of the distribution. The quartiles together with the median are often used by trade union negotiators. A negotiator would want to compare the median income of his members in a company with the median income of employees in comparable jobs. Similarly by using the lower quartile and the upper quartile a negotiator could compare incomes at the lower and upper ranges.

The median is sometimes designated Q_2.

Interquartile range

The difference $Q_3 - Q_1$ is called the *interquartile range*. In our example on company profits the interquartile range is $134 - 33 = 101$. Thus the spread of profit of the middle 50% of companies is £101,000.

The range

The difference between the largest and the smallest item in a set of observations is called the *range*. In our example on company profits the largest profit is 1 604 and the smallest profit 6; thus the range of profits is $1\ 604 - 6 = 1\ 598$. The range of profits of the companies is therefore £1 598 000. The range is used in quality control, but outside this field it has limited use.

Quartile deviation

The quartile deviation is the average of the differences of the quartiles from the median. In the case of the lower quartile this difference is $M - Q_1$ (where M is the median); for the upper quartile this difference is $Q_3 - M$.

The average of these differences is

$$\frac{(M - Q_1) + (Q_3 - M)}{2} = \frac{M - Q_1 + Q_3 - M}{2}$$

$$= \frac{Q_3 - Q_1}{2}$$

$$\text{Quartile deviation} = \frac{Q_3 - Q_1}{2}$$

The quartile deviation is half the interquartile range $Q_3 - Q_1$
In our example on companies the quartile deviation is given by:

$$\text{Quartile deviation} = \frac{134 - 33}{2} = \frac{101}{2} = 50.5$$

Quartile deviation is £50 500.

Exercise 2.2 (A)

For the data of exercise 2.2 find the lower quartile, upper quartile, interquartile range, range and the quartile deviation.

Deciles and percentiles

The *deciles* divide the distribution into ten equal parts and are usually denoted by $D_1, D_2, D_3 \ldots\ldots\ldots D_9$. D_1 is the lowest decile, D_9 the highest decile and D_5 is of course the median. D_1 is the value such that nine tenths of the observations are greater than D_1 and one tenth of the observations are less than D_1.

In income distributions those earning an income of D_1 or less are among the low paid. The Department of Employment in its published statistics (*New Earnings Survey*) gives D_1, Q_1, median, Q_3 and D_9 for most occupations; thus the spread of incomes may be determined.

The *percentiles* divide the distribution into one hundred equal parts. If an examination board were to offer a prize to the top 4% of the candidates, then the pass mark for the prize would be at the 96th percentile.

The calculation of deciles and percentiles requires more values than in the example given on page 32. The calculation of a decile is dealt with subsequently in the chapter.

Estimates of median, quartiles and deciles when data are given as a grouped frequency distribution.

Worked Example No. 2.3

Income of full time women employed in professional, technical, administrative, managerial and teaching occupations in 1971. Data obtained from *Family Expenditure Survey* 1971.

Range of weekly earnings	*No. of women*
Under £10	4
£10 and under £15	46
£15 and under £20	40
£20 and under £25	61
£25 and under £30	30
£30 and under £35	30
£35 and under £40	16
£40 and under £45	9
£45 and under £50	3
£50 and under £55	3
£55 and over	6
Total	248

The first step is to put the data into a cumulative distribution. We have to consider the boundaries between the ranges. £10 and under £15 means women earning between £10 and £14.99 inclusive (most accounting systems ignore half pence). Strictly speaking, the boundary should be at £14.99½, i.e. the mid point between greatest value of one range (£14.99) and the lowest value of next range (£15). In some situations we must give the actual boundaries, but as we have income ranges of £5 the difference between £14.99½ and £15 is so small in relation to £5 that it may be ignored. Thus, in this example, boundaries are taken at £10, £15, £20 . . . etc. rather than the strictly correct £9.99½, £14.99½, £19.99½.

To obtain a cumulative distribution we find the number earning less than £10; less than £15 etc. The number earning less than £15 is 4 + 46 = 50. The number earning less than £20 is 4 + 46 + 40 = 90. The number earning less than £25 is 4 + 46 + 40 + 61 = 151. The simplest way to obtain the cumulative frequency is to add the next frequency to the previous cumulative total. Thus the number earning less than £30 is 30 + 151 = 181.

	Cumulative frequency
Less than £10	4
Less than £15	50
Less than £20	90
Less than £25	151
Less than £30	181
Less than £35	211
Less than £40	227
Less than £45	236

Less than £50	239
Less than £55	242
Less than £80*	248†

* £80 chosen arbitarily but upper limit must be realistic.

† This figure must equal the total number of women shown in the frequency table. If the final cumulative total had not been 248, this would have meant that an error had occurred and it is then necessary to check the arithmetical accuracy.

In our example on company profits, there were 15 companies and the median was the eighth item. Now $8 = \frac{15 + 1}{2}$. Thus if we have n items the median occurs at the $\frac{(n + 1)}{2}$ item. However when n is large, as in the example on women's earnings, we take this to be $\frac{n}{2}$. Similarly the lower quartile occurs at the $\frac{(n + 1)}{4}$ item and upper quartile at the $\frac{3(n + 1)}{4}$ item, but again we take the values $\frac{n}{4}, \frac{3n}{4}$ when n is large.

Important note

If we are given data in percentage form, it is incorrect to use $\frac{(n + 1)}{2}$; we must use $\frac{n}{2}$ in the case of the median, and $\frac{n}{4}$ and $\frac{3n}{4}$ for the lower and upper quartiles respectively.

We can find the medians and quartiles either by a graphical method or by calculation. In examination questions you are often directed to use a particular method, for example the question may state 'Calculate the median'. In these circumstances you have to calculate. If the question merely states 'Find the median' or 'Estimate the median', you have a choice. However, you need to learn both graphical and calculation methods.

Estimation of medians, quartiles and deciles by a graphical method

The first step is to use the cumulative frequency table to draw the cumulative frequency curve (this curve is also called an *ogive*). Always plot the cumulative frequencies on the vertical axis; income is plotted on the horizontal axis. Corresponding to £10 we plot 4, to £15 we plot 50, to £20 we plot 90 and so on. Always label the axes, give the source of the data if this is supplied and give the graph a title — many examiners deduct a mark or two if this is not done.

Median occurs at $\frac{n}{2}$ item i.e. $\frac{248}{2} = 124^{th}$ item

Lower quartile occurs at $\frac{n}{4}$ item i.e. $\frac{248}{4} = 62^{nd}$ item

Upper quartile occurs at $\frac{3n}{4}$ item i.e. $\frac{3 \times 248}{4} = 186^{th}$ item

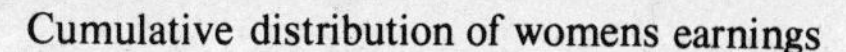

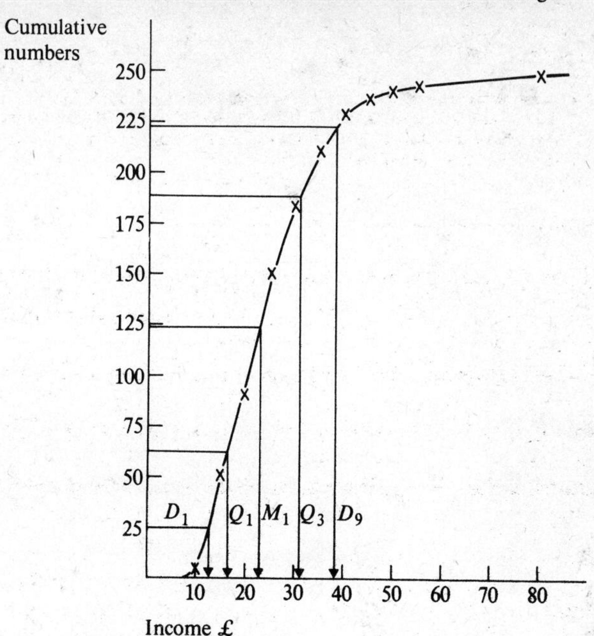

Figure 2.2 Source: *Family expenditure survey 1971*

[as a check $\frac{n}{4} + \frac{n}{2} = \frac{3n}{4}$, i.e. 62 + 124 = 186]

We draw horizontal lines from 124, 62 and 186 on the cumulative (vertical) axis until we meet the curve and drop verticals until we meet the income axis. The point where the verticals cut the income axis give the median income (about £22.80), lower quartile (about £16.70) and the upper quartile (about £30.50).

The quartile deviation is $\frac{Q_3 - Q_1}{2}$ and from data above

$$\frac{Q_3 - Q_1}{2} \text{ is } \frac{30.5 - 16.7}{2} = 6.9$$

We may also estimate the deciles by a graphical method. The position of the lowest decile is at the $\frac{n}{10}$ item and the position of the highest decile is at the $\frac{9n}{10}$ item. In this example $\frac{n}{10} = \frac{248}{10} = 24.8$ and $\frac{9n}{10} = \frac{9 \times 248}{10} = 223.2$.

We proceed in the same way and we find the lowest decile (about £13) and highest decile (about £38.20).

Similarly we could find the other deciles.

We can use the ogive to find the percentage of women who earn between £24 and £36 per week. In Figure 2.3 the ogive of Figure 2.2 is plotted again (in the examination room to save time, you can use the same ogive). We draw verticals from £24 and £36 until they meet the curve, then horizontal lines are drawn to meet the cumulative axis. We read off the numbers 215 and 140. Thus 215 – 140 = 75 women earn between £24 and £36. Expressed as a percentage of the total number of women, this is $\frac{75}{248} \times 100 = 30\%$

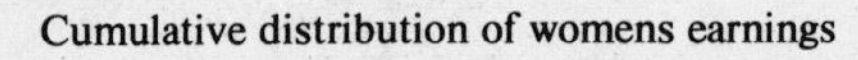

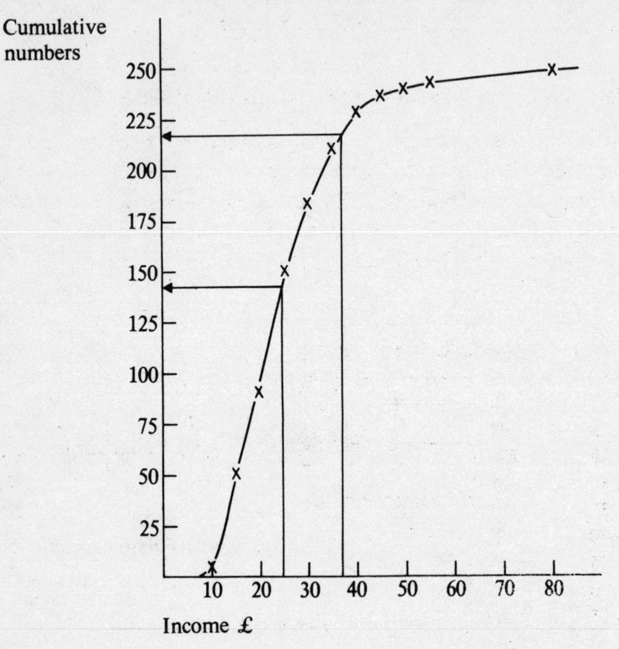

Figure 2.3 Source: *Family expenditure survey 1971*

Exercise 2.3

Number of miles covered in one day by 150 delivery vans:

Number of miles	*Number of vans*
under 30	3
30 and under 40	14
40 and under 50	30
50 and under 60	34
60 and under 70	28
70 and under 80	23
80 and under 90	15
90 and over	3
Total	150

i Draw an ogive (cumulative frequency curve) of the above data.
ii From your ogive find: (a) median; (b) lower quartile; (c) upper

quartile; (d) quartile deviation; (e) lowest decile; (f) highest decile.

iii Estimate the percentage of vans that cover between 48 and 77 miles per day.

Estimation of median, quartiles and deciles by calculation

The calculation method is based on simple proportion. Many students prefer to use a formula when using the calculation method. In these circumstances the method using the formula is explained first. Some students, however, prefer to understand the rationale of the method. This is explained at the end of the section.

The calculation is usually set out as below:

	Cumulative frequency	
Less than 10	4	← D_1
Less than 15	50	← Q_1
Less than 20	90	
		← M
Less than 25	151	
Less than 30	181	
		← Q_3
Less than 35	211	
Less than 40	227	
Less than 45	236	
Less than 50	239	
Less than 55	242	
Less than 80	248	

Median at position $\frac{n}{2} = \frac{248}{2} = 124$

Q_1 at position $\frac{n}{4} = \frac{248}{4} = 62$

Q_3 at position $\frac{3n}{4} = 3 \times \frac{248}{4} = 186$

D_1 at position $\frac{n}{10} = \frac{248}{10} = 24.8$

The next step is to insert arrows in the cumulative frequency table. As the median is at position 124, insert an arrow marked M between 90 and 151. As Q_1 is at position 62, insert an arrow marked Q_1 between 50 and 90, and so on.

The formula for the median is

$$\text{Median} = L_1 + (L_2 - L_1)\left(\frac{n/2 - \Sigma f_1}{\Sigma f_2 - \Sigma f_1}\right)$$

where L_1 = lower class boundary of median class

L_2 = upper class boundary of median class

n = total number of items

Σf_1 = cumulative frequency immediately less than the median

Σf_2 = cumulative frequency immediately greater than the median

In order to find L_1, L_2, Σf_1, Σf_2 it is easiest to refer to the arrow marked M in the table above. L_1 and Σf_1 are the values above the arrow, L_2 and Σf_2 are below the arrow.

$$L_1 = 20;\ L_2 = 25;\ \Sigma f_1 = 90;\ \Sigma f_2 = 151;\ n = 248.$$

$$\text{Median} = 20 + (25 - 20)\left(\frac{248/2 - 90}{151 - 90}\right)$$

$$= 20 + 5 \times \frac{34}{61}$$

$$= 20 + 2.79 = £22.79$$

Median is at £22.80 (to the nearest 10p)

The formulae for the lower quartile and upper quartile are very similar to that for the median:

$$Q_1 = L_1 + (L_2 - L_1)\left(\frac{n/4 - \Sigma f_1}{\Sigma f_2 - \Sigma f_1}\right)$$

Where L_1, L_2, Σf_1, Σf_2 relate to the lower quartile instead of the median, i.e. $L_1 = 15$, $L_2 = 20$, $\Sigma f_1 = 50$, $\Sigma f_2 = 90$, $n = 248$

$$\text{then } Q_1 = 15 + (20 - 15)\left(\frac{248/4 - 50}{90 - 50}\right)$$

$$= 15 + \frac{5 \times (62 - 50)}{90 - 50}$$

$$= 15 + \frac{5 \times 12}{40}$$

$$= 15 + 1.5 = £16.50$$

$$Q_3 = L_1 + (L_2 - L_1)\left(\frac{3n/4 - \Sigma f_1}{\Sigma f_2 - \Sigma f_1}\right)$$

Here $L_1 = 30$, $L_2 = 35$, $\Sigma f_1 = 181$, $\Sigma f_2 = 211$, $n = 248$

$$Q_3 = 30 + (35 - 30)\frac{\left(\frac{3 \times 248}{4} - 181\right)}{211 - 181}$$

$$= 30 + 5 \times \left(\frac{186 - 181}{211 - 181}\right)$$

$$= 30 + \frac{5 \times 5}{30} = 30 + 0.83 = 30.83$$

$$= £30.80 \text{ (to the nearest 10p)}$$

Similarly the deciles may be calculated — the method will be illustrated by calculating D_1:

$$D_1 = L_1 + (L_2 - L_1)\left(\frac{\frac{n}{10} - \Sigma f_1}{\Sigma f_2 - \Sigma f_1}\right)$$

$$L_1 = 10,\ L_2 = 15,\ \Sigma f_1 = 4,\ \Sigma f_2 = 50,\ n = 248$$

$$D_1 = 10 + (15 - 10)\left(\frac{24.8 - 4}{50 - 4}\right)$$

$$= 10 + 5 \times \frac{20.8}{46}$$

$$= 10 + 2.26$$

$$= 12.26$$

$$= £12.30 \text{ (to the nearest 10p)}$$

It will be noted that the graphical method and calculation method give very similar results, the major divergence occurring at D_1 — the graphical method giving £13 compared with £12.30 by calculation. The calculation method is based on simple proportion which assumes the points on the graph are joined by straight lines; the graphical method uses a smooth curve. For Q_1, median and Q_3 the curve and the straight line joining the points are clearly close together, but in the case of D_1 the curve and the straight line joining the two points diverge and this accounts for the differing results.

Calculation of the median without using the formula

The calculation method is based on simple proportion; using the example on women's income the median is at the position of the 124th item.

4	women have an income less than	£10
50	women have an income less than	£15
90	women have an income less than	£20
124	women have an income less than	M
151	women have an income less than	£25
181	women have an income less than	£30

The median is at the 124th item and we assume it divides the range 20–25 in the same proportion as 124 divides 90 to 151.

i.e. $$\frac{M-20}{124-90} = \frac{25-20}{151-90}$$

$$\frac{M-20}{34} = \frac{5}{61}$$

$$M-20 = \frac{5 \times 34}{61}$$

$$M = 20 + \frac{5 \times 34}{61}$$

$$M = 20 + 2.79 = £22.79$$

∴ Median is at £22.80 (to the nearest 10p)

Exercise 2.3A

Using the data of Exercise 2.3, calculate (a) median; (b) lower quartile; (c) upper quartile; (d) quartile deviation; (e) lowest decile; (f) highest decile.

Worked Example No. 2.4

In worked example no. 2.3 the class intervals were always £5 and the plotting of the ogive was reasonably straightforward. When the class intervals differ the graphical method needs to be considered carefully.

Men wholly unemployed in Great Britain in January 1972 — Analysis of duration of unemployment

Duration in weeks	*Number of men (thousands)*
1 or less	47
Over 1 up to 2	44
Over 2 up to 3	20
Over 3 up to 4	29
Over 4 up to 5	28
Over 5 up to 6	29
Over 6 up to 7	26
Over 7 up to 8	23
Over 8 up to 13	105
Over 13 up to 26	147
Over 26 up to 39	77
Over 39 up to 52	42
Over 52	129
Total	746

Source: *Department of Employment Gazette*

i Use a graphical method to find the (a) median; (b) lower quartile; (c) upper quartile; (d) lowest decile.

ii Calculate the median and thus check your answer for the median found in part (i).

i Graphical method

Position of:

$$D_1 = \frac{n}{10} = \frac{746}{10} = 74.6$$

$$Q_1 = \frac{n}{4} = \frac{746}{4} = 186.5$$

$$\text{Median} = \frac{n}{2} = \frac{746}{2} = 373$$

$$Q_3 = \frac{3n}{4} = 3 \times \frac{746}{4} = 559.5$$

Less than 1	47	
Less than 2	91	
Less than 3	111	
Less than 4	140	
Less than 5	168	
Less than 6	197	
Less than 7	223	
Less than 8	246	
Less than 13	351	
Less than 26	498	← *M*
Less than 39	575	
Less than 52	617	
Less than 104*	746	

* Upper range of table taken as 104, even though some men who are 'unemployable' will have been on register many years

It will be noted that the position of the upper quartile is less than 39 weeks. If we plotted all the data as in Figure 2.4, the scale is small at the relevant points. If we plot up to 39 weeks only as in Figure 2.5, we have a larger scale and can thus obtain greater accuracy.

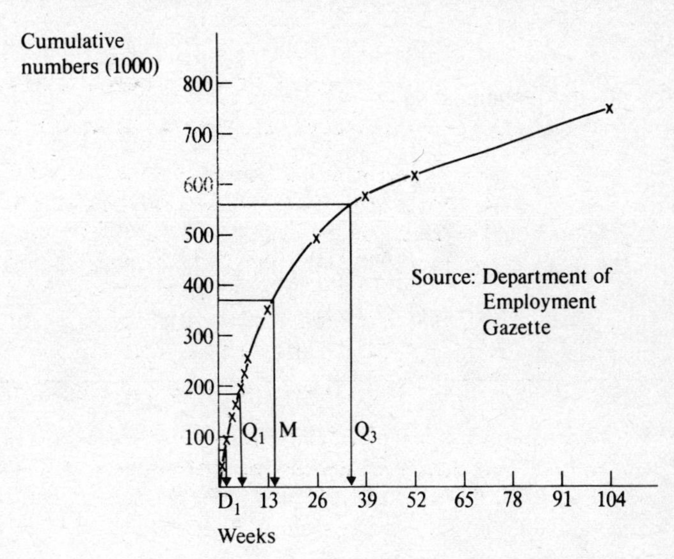

Figure 2.5 Source: *Department of Employment Gazette*

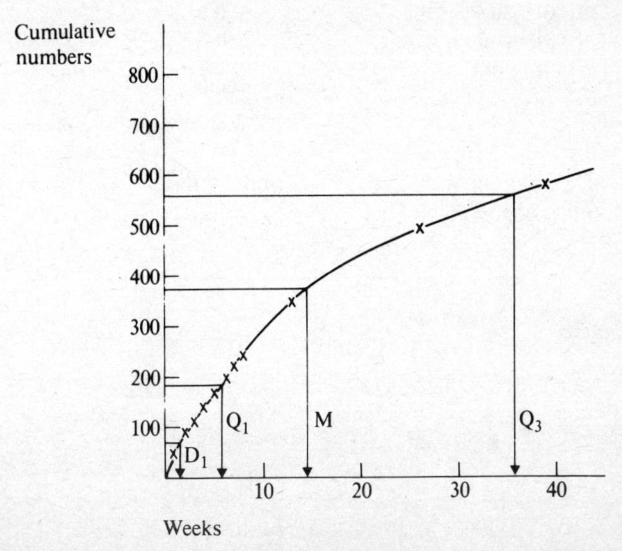

Figure 2.4 Source: *Department of Employment Gazette*

Note that in both Figures 2.4 and 2.5 the horizontal axis has a constant scale, i.e. in Figure 2.4 the horizontal difference between 0 and 13 weeks is the same as the distance between 26 and 39 weeks. If a constant scale is not used, marks are usually lost.

It is clear that we can read off the results more accurately from Figure 2.5. We find $D_1 = 1\frac{1}{2}$ weeks, $Q_1 = 5\frac{3}{4}$ weeks, $M = 14\frac{1}{2}$ weeks $Q_3 = 35\frac{3}{4}$ weeks.

ii Calculation method

$$M = L_1 + (L_2 - L_1)\left(\frac{\frac{n}{2} - \Sigma f_1}{\Sigma f_2 - \Sigma f_1}\right)$$

$L_1 = 13, L_2 = 26, \Sigma f_1 = 351, \Sigma f_2 = 498, n = 746$

$$M = 13 + (26-13)\left(\frac{373 - 351}{498 - 351}\right) = 13 + \frac{13 \times 22}{147} = 13 + 1.945$$

$= 14.945$ weeks (14.9 weeks)

Exercise 2.4

Size of shareholding	*Percentage of shareholders*
1 - 100	13.4
101 - 500	51.7
501 - 1 000	19.2
1 001 - 2 000	9.5
2 001 - 5 000	4.0
5 001 - 10 000	0.9
10 001 - 50 000	0.9
50 001 and over	0.4
Total	100.0

Source: *Company accounts 1976*

Use a graphical method to find (i) median; (ii) lower quartile; (iii) upper quartile.

Important note

Examination questions on medians and quartiles are usually well done. Marks are, however, frequently lost by carelessness. The following notes should help reduce this loss of marks.

1 Check that the total frequency and cumulative total are the same.
2 Consider the boundaries of class intervals carefully. In particular, in the case of ages, 25–29 means 25 but under 30.

In graphical work:

3 Give the graph a title and state source of data — both axes should be clearly labelled.

4 Use a constant scale; if the class interval changes, the horizontal distance between your plots changes.

5 In graphical work be neat and tidy. The graph should be drawn with a 2H pencil or at least a pencil with a sharp point!

6 If you attempt a calculation and a graphical method your answers are likely to be slightly different. If your answers diverge considerably one (or both) of your methods is wrong.

Finally give the units with your answer. Some examiners insist on (say) 14.9 weeks for full marks. 14.9 without units could lose you a mark.

Revision Exercises

2.5 **a** Give the relative advantages and disadvantages of the mean, median and mode as measures of location.

b Calculate these measures for the following data:

28, 9, 14, 19, 3, 33, 22, 7, 14, 26, 10, 16, 20

2.6 Define the terms: mean, median, mode, range.

Illustrate the steps involved in calculating these values by estimating each of them for the following set of data:

15, 14, 17, 18, 16, 17, 20, 16, 15, 13, 16, 19, 11, 16, 18, 15

(I.O.S. 1971)

2.7 **Intelligence quotients (I.Q.) of 100 children at a junior school**

I.Q.	*Number of children with given I.Q.*
50 – 59	1
60 – 69	2
70 – 79	8
80 – 89	18
90 – 99	23
100 – 109	21
110 – 119	15
120 – 129	9
130 – 139	3

Calculate (a) the mode, (b) the median of the above data. Check your results by a graphical method.

2.8 Age of U.K. merchant vessels

Age	*Tankers* (*No.*)	*Passenger vessels* (*No.*)
Under 5 years	137	18
5 - 10 years	113	16
10 - 15 years	151	27
15 - 20 years	62	20
20 - 25 years	15	31
25 years and over	8	17
	486	129

Compare age distributions of tankers and passenger vessels by calculating for each distribution the median and upper quartile.
Comment on your results.

(H.N.D. 1976)

2.9 Age distribution of dangerous drug addicts in the U.K. — 1971 and 1974

Age distribution	*Number of Addicts*	
	1971	*1974*
under 20 years	118	64
20 and under 25 years	722	694
25 and under 30 years	288	686
30 and under 35 years	112	163
35 and under 50 years	112	165
50 years and over	177	200
Age not stated	20	8

Source: *Social Trends*

a Draw cumulative frequency curves (ogives) for the above data and use these curves to find for each year: (i) median; (ii) lower quartile; and (iii) upper quartile, of the age of the addicts. Hence find the quartile deviation for each year.

b Comment on your results.

(I.O.S. May 1977)

2.10 **a** Briefly explain why medians and quartiles are often used as measures of 'average' and dispersion in income distribution.

Income of households in Greater London area

Range of income per week	*Percentage of households*
Under £10	2
£10 but under £20	15
£20 but under £30	10

Range of income per week	*Percentage of households*
£30 but under £40	9
£40 but under £50	11
£50 but under £60	12
£60 but under £70	11
£70 but under £80	8
£80 but under £100	11
£100 and over	11

Source: *Family Expenditure Survey* 1974

b Estimate the median, quartiles, quartile deviation and highest decile for the above distribution.

c The corresponding values for Scotland were median £46.20, lower quartile £27.30, upper quartile £68.20, quartile deviation £20.50, highest decile £95. Compare the two distributions and comment on your results.

(B.A.S.S. (Part Time) 1976)

2.11 Define **a** median; **b** quartiles; **c** quartile deviation. Calculate the median, quartiles and quartile deviation for the sample of fifteen values given below:

9, 0, 4, 5, 1, 10, 8, 3, 7, 0, 9, 4, 7, 2, 6

2.12 *Answer all parts*
The following information relates to the ages of unemployed males in Scotland and East Anglia, July 1977.

Age	*Scotland* *Cumulative % freq.*	*East Anglia* *Cumulative % freq.*
Under 20	22.8	19.8
Under 30	50.1	44.1
Under 40	66.8	59.0
Under 50	79.8	70.1
Under 60	91.2	83.8
Under 70	100.0	100.0

i Sketch the graphs of the cumulative percentage frequency distributions.

ii Estimate the median ages for each distribution from your graph.

iii The mean values are, for Scotland, 34.6 years and for East Anglia, 37.9 years. What percentage of unemployed males in each distribution are older than the respective means?

iv Use your graph and estimate a measure of dispersion for each distribution.

v Give a written comparison of the distributions based on your above results.

(B.A.S.S. (Full Time) 1978)

2.13

Duration of unemployment in weeks	*No. of women*
2 or less	1045
Over 2 up to 5	1201
Over 5 up to 8	1032
Over 8 up to 13	1252
Over 13 up to 26	1034
Over 26 up to 52	657
Over 52	962
	7183

(a) Calculate from the above distribution the median and quartile deviation.

(b) Represent the data graphically and estimate the median from your graph.

(c) Also estimate from your graph the number of women who had been unemployed for a period between 11 and 16 weeks.

(B.Sc. (Soc.) 1962)

3 Mean and Standard Deviation

Mean

The mean of a series of values is simply the arithmetic average. The mean was found in the chapter on median and quartiles, and is obtained by adding the values and then dividing by the number of values. We use again the example from Chapter 2.

Worked Example No. 3.1.

Profits of 15 Companies (in thousands of pounds)

27, 6, 38, 102, 54, 39, 70, 33, 73, 152, 1604, 380, 134, 46, 17

The mean profit of these companies is
(27 + 6 + 38 + 102 + 54 + 39 + 70 + 33 + 73 + 152 + 1604 + 380 + 134 + 46 + 17) ÷ 15

$$= \frac{2\,775}{15} = 185$$

Exercise 3.1

The following data give the sales commission of 10 representatives employed by a firm. Find the mean sales of the representatives.

41, 72, 31, 84, 54, 52, 60, 49, 43, 64

In mathematical terms the mean is given by

$$\overline{x} = \frac{\Sigma x}{n}$$

where n is the number of observations. Σ is the mathematical sign which means 'sum'.

Calculation of mean when data is in a frequency table

Worked Example No. 3.2

Children in households of recidivist prisoners

Number of children in the household	*Number of households*
None	25
1	40
2	42

Number of children in the household	*Number of households*
3	28
4	18
5	11
6	9
7 or more	6
Total	179

Source: Pauline Morris, *Prisoners and their Families.*

Calculate the mean number of children per household.

In this example we need to find the total number of children and then to find the mean we divide by the number of households.

The number of children is $(1 \times 40) + (2 \times 42) + (3 \times 28) + (4 \times 18) + (5 \times 11) + (6 \times 9) + (8 \times 6) = 437$

In this calculation we have assumed that the mean number of children in the group shown as 7 or more is 8.

The total number of households is 179

Hence mean number of children per household is $\dfrac{437}{179}$

$= 2.441$ children.

The actual calculation is simplified if the following tabulation is employed:

x	f	fx
0	25	0
1	40	40
2	42	84
3	28	84
4	18	72
5	11	55
6	9	54
8	6	48
Total	179 ($= \Sigma f$)	437 ($= \Sigma fx$)

The mean is then given by the formula

$$\bar{x} = \frac{\Sigma fx}{\Sigma f} = \frac{437}{179} = 2.441 \text{ children}$$

Important note

When candidates are not allowed to take calculators into the examination room, it is essential to learn the method of finding a mean and standard deviation using an assumed mean. When candidates are allowed to take a calculator into the examination room, students have a choice. With an assumed mean the numbers are smaller but the formulae and the procedure are more complicated. If an assumed mean is not used, the procedure is simpler but the numbers are larger. In general, students who are weak on numerical work are more successful if they *do not use* an assumed mean. If students have an 'overflow' problem on calculators because of large numbers, the method of dealing with this situation is explained on page 59.

Use of an assumed mean

The calculations can often be simplified by the use of an assumed mean.

The most convenient assumed mean is usually the value of x corresponding to the largest frequency. In the example above the largest frequency is 42 and the corresponding value of x is 2.

The method is illustrated in the tabulation below:

x	f	$x_1 = (x - 2)$	fx_1	
0	25	-2	-50	
1	40	-1	-40	
2	42	0		
3	28	1	28	
4	18	2	36	
5	11	3	33	
6	9	4	36	
8	6	6	36	
Total	179 (= Σf)		-90 + 169 = 79	(= Σfx_1)

The formulae for the mean then becomes

$$\bar{x} = A + \frac{\Sigma fx_1}{\Sigma f}$$

where A is the assumed mean, i.e. $A = 2$

$$\bar{x} = 2 + \frac{79}{179} = 2 + 0.441 = 2.441 \text{ children}$$

Exercise 3.2

Number of rooms in dwellings in Great Britain in 1975

No. of rooms	*Percentage of dwellings*
1	1
2	2
3	8
4	21
5	33
6	26
7	5
8 or more	4
Total	100

Calculate the mean number of rooms per dwelling in Great Britain.

Worked Example No. 3.3

Mother's age at birth of child for a sample of mothers

Mother's age	*No.*
Under 17	1
17 - 19	33
20 - 24	287
25 - 29	367
30 - 34	250
35 - 39	134
40 - 44	41
Over 44	8
Not stated	26
	1 142

Calculate the arithmetic mean.

In the calculation of the mean it is necessary to make some assumptions about the 'open-ended' classes, firstly 'under 17'. What is the average age of mothers in this class? The estimate is of necessity arbitrary but must be realistic. Some students have given 8 or $8\frac{1}{2}$ years but this is impossible. Have you ever heard of a mother aged 8? In the calculation the average age for this group is taken as 16.5 years. Secondly, over 44 years; in this calculation the average age for this group is taken as 47.5 years, but this may well be rather high.

The 26 mothers whose age is not stated have been completely ignored from the calculation.

For other groups the average age is taken as the midpoint i.e. for 20 - 24 the average is taken to be 22.5 since 20 - 24 means people aged from 20 years to 24 years and 364 days, and the average of this group is 22.5 years.

Mother's age	*Number* (f)	*Central value* (x) (x)	fx
Under 17	1	16.5	16.5
17 - 19	33	18.5	610.5
20 - 24	282	22.5	6 345.0
25 - 29	367	27.5	10 092.5
30 - 34	250	32.5	8 125.0
35 - 39	134	37.5	5 025.0
40 - 44	41	42.5	1 742.5
Over 44	8	47.5	380.0
Total	1 116		32 337.0

$$\bar{x} = \frac{\Sigma fx}{\Sigma f} = \frac{32\ 337.0}{1\ 116} = 28.98 \text{ years}$$

Use of an assumed mean

This problem can also be solved using an assumed mean.

In the tabulation below an assumed mean is taken to be 27.5 and a class interval of 5 is used to simplify the arithmetic (it is not essential to use a class interval; in some cases a class interval is a considerable advantage, in other cases it may be even a disadvantage).

Mother's Age	*Number* f	*Central value* x	*Central value* (A = 27.5) $x_1 = x - A$	(C = 5) $\frac{x - A}{C} = x_2$	fx_2
Under 17	1	16.5	-11	-2.2	- 2.2
17 - 19	33	18.5	- 9	-1.8	- 59.4
20 - 24	282	22.5	- 5	-1	-282.0
25 - 29	367	27.5	0	0	
30 - 34	250	32.5	5	+1	+250
35 - 39	134	37.5	10	+2	+268
40 - 44	41	42.5	15	+3	+123
Over 44	8	47.5	20	+4	+ 32
	1 116				-343.6 + 673 = 329.4

The formula for the assumed mean now becomes

$\bar{x} = A + C \times \frac{\Sigma fx_2}{\Sigma f}$ where C is the class interval

In this example A = 27.5; C = 5

$$\bar{x} = 27.5 + 5 \times \frac{329.4}{1\ 116} = 27.5 + 1.48 = 28.98 \text{ years}$$

N.B. Check that your answer is realistic. A student once gave an answer to this question of 369 years! The student had taken the assumed mean to be the frequency (367) instead of corresponding central value 27.5.

Exercise 3.3

Sales (£1 000)	*Number of representatives*
Under 10	3
10 and under 20	8
20 and under 30	16
30 and under 40	13
40 and under 50	7
50 and over	3

Calculate the mean of the above distribution.

Measures of dispersion

In Chapter 2 it was mentioned that the quartiles and the deciles measure the spread of the distribution. These measures are particularly useful for distributions which are not symmetrical, e.g. income distributions. There are two measures of dispersion based on the mean: (i) *mean deviation*; and (ii) *standard deviation*. The important measure is the standard deviation. The mean deviation is hardly ever used in practice but examiners very occasionally set a question which involves mean deviation.

Mean deviation

The mean deviation is found by taking the arithmetic difference between each observation and the mean. The arithmetic difference means that the sign is disregarded. These differences are totalled and then divided by the number of observations. This may be expressed in mathematical terms as where x is an observation $|x - \bar{x}|$ means the arithmetic difference. The sum of these arithmetic differences is $\Sigma|x - \bar{x}|$ and mean deviation is given by

$$\text{Mean deviation} = \frac{\Sigma|x - \bar{x}|}{n}$$

Worked Example No. 3.4

Using the data of worked example no. 3.1 find the mean deviation.

The mean was found to be 185. The difference between the first observation 27 and the mean is $|x - \bar{x}| = |27 - 185| = 158$; for the second observation $|6 - 185| = 179$. The differences for the third and subsequent observations are found to be 147, 83, 131, 146, 115, 152, 112, 33, 1419, 195, 51, 139, 168

The mean deviation is given by

$$(158 + 179 + 147 + 83 + 131 + 146 + 115 + 152 + 112 + 33 + 1419 + 195 + 51 + 139 + 168) \div 15$$

$$= \frac{3\,228}{15} = 215.2 \quad (£215\,200)$$

Exercise 3.4

Find the mean deviation for the data for Exercise 3.1.

Standard deviation

The mean deviation presents some theoretical problems because we do not know the sign of $x - \bar{x}$ in the general case of an observation x. The standard deviation gets over this problem by squaring $x - \bar{x}$. $(x - \bar{x})^2$ is positive whether or not $x - \bar{x}$ is negative or positive.

The standard deviation is found by taking the difference between the observation and the arithmetic mean i.e. in mathematical terms $x - \bar{x}$, squaring i.e. $(x - \bar{x})^2$, adding all the observations i.e. $\Sigma(x - \bar{x})^2$, taking the average i.e. $\frac{\Sigma(x - \bar{x})^2}{n}$, and then taking the square root i.e. $\sqrt{\frac{\Sigma(x - \bar{x})^2}{n}}$

i.e. Standard deviation $\sigma = \sqrt{\frac{\Sigma(x - \bar{x})^2}{n}}$

If we have grouped data, the formula for standard deviation becomes

$$\sigma = \sqrt{\frac{\Sigma f(x - \bar{x})^2}{\Sigma f}}$$

However, **it is not recommended that the above formula be used to find the standard deviation** of a grouped frequency distribution, since it may be shown that the above formula is equal to

$$\sigma = \sqrt{\frac{\Sigma fx^2}{\Sigma f} - \left(\frac{\Sigma fx}{\Sigma f}\right)^2}$$

and this is the formula that should be used.

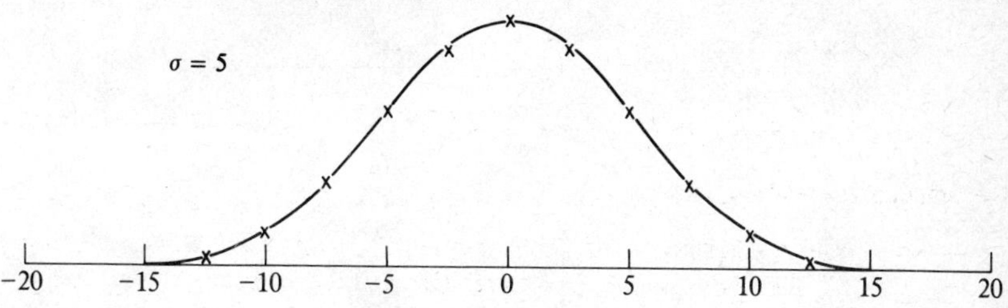

Figure 3.1

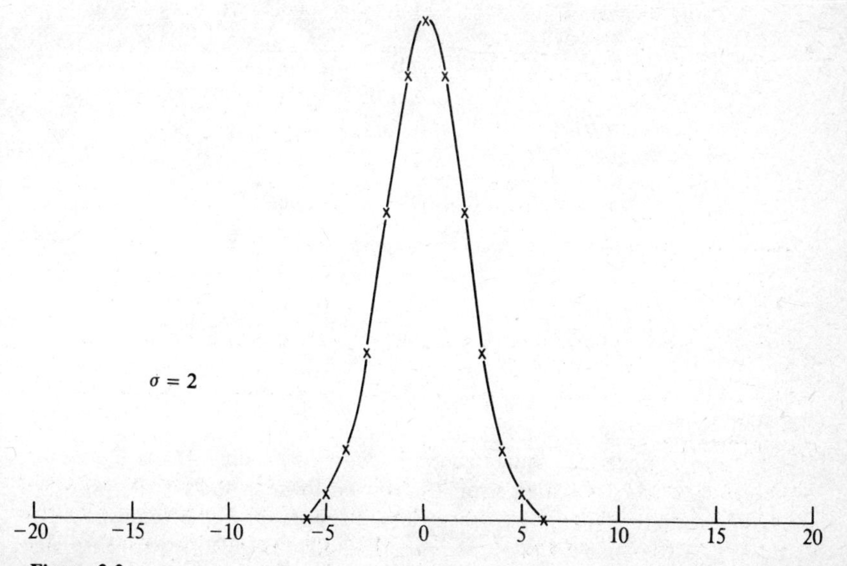

Figure 3.2

σ measures the spread of a distribution. Large values of σ show that observations are spread widely about the mean. Small values of σ show that observations are close to the mean.

Worked Example No. 3.5

Suppose we now wish to find the standard deviation of the data of worked example no. 3.2. From the formula

$$\sigma = \sqrt{\frac{\Sigma fx^2}{\Sigma f} - \left(\frac{\Sigma fx}{\Sigma f}\right)^2}$$

we see that we need to find Σfx^2, Σfx, Σf. In worked example 3.2 we have already found Σf and Σfx, so we now need to add an extra column to enable us to find Σfx^2.

x	f	fx	fx^2
0	25	0	0
1	40	40	40
2	42	84	168
3	28	84	252
4	18	72	288
5	11	55	275
6	9	54	324
8	6	48	384
	179 (= Σf)	437 (= Σfx)	1 731 (= Σfx)^2

$$\sigma = \sqrt{\frac{\Sigma fx^2}{\Sigma f} - \left(\frac{\Sigma fx}{\Sigma f}\right)^2} = \sqrt{\frac{1\,731}{179} - \left(\frac{437}{179}\right)^2}$$

$$= \sqrt{9.6704 - 5.9601} = \sqrt{3.7103}$$

$$= 1.926 \text{ children}$$

N.B.: It is convenient to use the same table to find $\bar{x}$:

$$\bar{x} = \frac{\Sigma fx}{\Sigma f} = \frac{437}{179} = 2.441 \text{ children}$$

Important notes

i The number under the square root sign *must be positive*. If the number is negative, you have made an error. Check your formula and your working.

ii Make certain that you have correctly labelled the 'x' and 'f' columns.

iii In the fourth column a very common error is to find $\Sigma(fx)^2$ instead of Σfx^2, i.e. students square the third column, fx. Remember fx^2 is f times x^2.

Use of an assumed mean

x	f	$x_1 = (x-2)$	fx_1	$f(x_1)^2$
0	25	−2	− 50	100
1	40	−1	− 40	40
2	42	0		
3	28	1	28	28
4	18	2	36	72
5	11	3	33	99
6	9	4	36	144
8	6	6	36	216
Total	179 (= Σf)		− 90 + 169 + 79 (= Σfx_1)	699 (= $\Sigma f(x_1)^2$)

$$\sigma = \sqrt{\frac{\Sigma f(x_1)^2}{\Sigma f} - \left(\frac{\Sigma fx_1}{\Sigma f}\right)^2} = \sqrt{\frac{699}{179} - \left(\frac{79}{179}\right)^2}$$

$$= \sqrt{3.9050 - 0.1947} = \sqrt{3.7103}$$

$$= 1.926 \text{ children}$$

N.B. It is convenient to use the same table to find the mean $\overline{x}$:

$$\overline{x} = A + \frac{\Sigma fx_1}{\Sigma f} = 2 + \frac{79}{179} = 2.441 \text{ children}$$

Exercise 3.5

Find the standard deviation for the data of Exercise 3.2.

Calculation of standard deviation when data is in a grouped frequency table

Worked Example No. 3.6

Suppose we now wish to find the standard deviation of the data of Worked Example No. 3.3.

Mother's age	*Number* f	*Central value* x	fx	fx^2
Under 17	1	16.5	16.5	272.25
17 - 19	33	18.5	610.5	11 294.25
20 - 24	282	22.5	6 345.0	142 762.50
25 - 29	367	27.5	10 092.5	277 543.75
30 - 34	250	32.5	8 125.0	264 062.50

Mother's age	*Number* f	*Central value* x	fx	fx^2
35 - 39	134	37.5	5 025.0	188 437.50
40 - 44	41	42.5	1 742.5	74 056.25
Over 44	8	47.5	380.0	18 050.00
	1 116		32 337.0	976 479.00

$$\bar{x} = \frac{\Sigma fx}{\Sigma f} = \frac{32\ 337}{1\ 116} = 28.98 \text{ years}$$

$$\sigma = \sqrt{\frac{\Sigma fx^2}{\Sigma f} - \left(\frac{\Sigma fx}{\Sigma f}\right)^2} = \sqrt{\frac{976\ 479}{1\ 116} - (28.98)^2}$$

$$= \sqrt{874.981 - 839.609} = \sqrt{35.372}$$

$$= 5.95 \text{ years}$$

Exercise 3.6

Find the standard deviation for the data of Exercise 3.3.

Methods for calculating mean and standard deviation when the central values are large

The examples so far employed were chosen so that students with inexpensive calculators would not have an 'overflow' problem i.e. numbers too large for the calculator to handle. There are, however, some exercises where some students will have an overflow problem and the following example illustrates how this problem can be tackled.

Worked Example No. 3.7

Gross annual earnings of males in full-time employment in 1975

Gross annual earnings (£)	*Percentage*
Under 1 000	3
1 000 and under 1 500	5
1 500 and under 2 000	4
2 000 and under 2 500	15
2 500 and under 3 000	22
3 000 and under 3 500	19
3 500 and under 4 000	14
4 000 and under 5 000	10
5 000 and under 6 000	2
6 000 and over	6

Source: *General Household Survey* 1975

Find the mean and standard deviation.

Method 1

This involves dividing the gross annual earnings by some convenient number (in this case 100 would be appropriate), then proceeding as in Worked Example No. 3.6. The final answers for $\bar{x}$ and σ are found by multiplying by 100.

Gross annual earnings (£100)	*Percentage* f	*Central value* x	fx	fx^2
Under 10	3	7.5	22.5	168.75
10 and under 15	5	12.5	62.5	781.25
15 and under 20	4	17.5	70.0	1 225.00
20 and under 25	15	22.5	337.5	7 593.75
25 and under 30	22	27.5	605.0	16 637.50
30 and under 35	19	32.5	617.5	20 068.75
35 and under 40	14	37.5	525.0	19 687.50
40 and under 50	10	45	450.0	20 250.00
50 and under 60	2	55	110.0	6 050.00
60 and over	6	75	450.0	33 750.00
	100		3 250.0	126 212.50

$$\bar{x} = \frac{\Sigma fx}{\Sigma f} = \frac{3\,250}{100} = 32.5$$

To convert back to original figures, multiply by 100:
Mean income = 32.5 × 100 = £3 250

$$\sigma = \sqrt{\frac{\Sigma fx^2}{\Sigma f} - \left(\frac{\Sigma fx}{\Sigma f}\right)^2} = \sqrt{\frac{126\,212.50}{100} - (32.5)^2}$$

$$= \sqrt{1\,262.125 - 1\,056.25} = \sqrt{205.875} = 14.35$$

Convert back to original figures by multiplying by 100:
Standard deviation = 14.35 × 100 = £1 435

Method 2

Use an assumed mean A and employ a constant interval C. The mean is then

$$\bar{x} = A + C\frac{\Sigma fx}{\Sigma f}$$

$$\text{and } \sigma = C\sqrt{\frac{\Sigma fx^2}{\Sigma f} - \left(\frac{\Sigma fx}{\Sigma f}\right)^2}$$

Gross annual earnings	*Percentage* f	*Central value* x'	(x=2 750) $x' - A$	(C=500) $\frac{x' - A}{C} = x$	fx	fx^2
Under 1 000	3	750	−2 000	−4	− 12	48
1 000 and under 1 500	5	1 250	−1 500	−3	− 15	45
1 500 and under 2 000	4	1 750	−1 000	−2	− 8	16
2 000 and under 2 500	15	2 250	− 500	−1	− 15	15

2 500 and under 3 000	22	2 750	– 0	0		
3 000 and under 3 500	19	3 250	+ 500	1	+ 19	19
3 500 and under 4 000	14	3 750	+1 000	2	+ 28	56
4 000 and under 5 000	10	4 500	+1 750	3.5	+ 35	122.5
5 000 and under 6 000	2	5 500	+2 750	5.5	+ 11	60.5
6 000 and over	6	7 500	+4 750	9.5	+ 57	541.5
	100				150 – 50 100	923.5

$$\bar{x} = A + C\frac{\Sigma fx}{\Sigma} = 2\,750 + 500 \times \frac{100}{100} = 2\,750 + 500$$

$$= £3\,250$$

$$\sigma = C\sqrt{\frac{\Sigma fx^2}{\Sigma f} - \left(\frac{\Sigma fx}{\Sigma f}\right)^2} = 500\sqrt{\frac{923.5}{100} - \left(\frac{100}{100}\right)^2}$$

$$= 500\sqrt{9.235 - 1} = 500\sqrt{8.235}$$

$$= 500 \times 2.8697 = £1\,435$$

Exercise 3.7

Annual income by highest educational qualification

Annual income £	*GCE 'O' level or CSE Grade 1*	*No qualification*
Less than 500	5	3
500 but less than 1 000	12	10
1 000 but less than 1 500	18	32
1 500 but less than 2 000	23	30
2 000 but less than 2 500	20	15
2 500 but less than 3 000	11	5
3 000 and over	11	5
	100	100

Source: *General Household Survey* 1973

For both distributions calculate the mean and standard deviation.

Important note

In most examinations for which this book is designed the calculation of the standard deviation is from a frequency table and Worked Examples 3.5, 3.6 & 3.7 cover most questions in this area. Occasionally examination questions are set where the data is given as individual values, e.g. as in Worked Example 3.1 where we were given the profits of 15 companies. In this case the formula to use is

$$\sigma = \sqrt{\frac{\Sigma(x - \bar{x})^2}{n}}$$

In more advanced books on statistics this formula is given in the technically correct form $\sigma = \sqrt{\frac{\Sigma(x - \bar{x})^2}{n - 1}}$. Most examiners accept either formula; however, some examiners insist on the technically correct formula where the divisor is $(n - 1)$ for full marks.

Worked Example No. 3.8

Find the standard deviation for the data of Worked Example no. 3.1.
In Worked Example no. 3.4 we found $|x - \bar{x}|$ for each observation.
We need $\Sigma(x - \bar{x})^2$ so we need to square each value $|x - \bar{x}|$ and add.

$$\text{Thus } \Sigma(x - \bar{x})^2 = 158^2 + 179^2 + 147^2 + \dots + 139^2 + 168^2$$
$$= 24\,964 + 32\,041 + 21\,609 + \dots + 19\,321 + 28\,224$$
$$= 2\,275\,674$$

$$\text{Using } \sqrt{\frac{\Sigma(x - \bar{x})^2}{n}} \text{ we have } \sigma = \sqrt{\frac{2\,275\,674}{15}}$$
$$= \sqrt{151\,711.6} = 389.50$$

$$\text{Using } \sqrt{\frac{\Sigma(x - \bar{x})^2}{n - 1}} \text{ we have } \sigma = \sqrt{\frac{2\,275\,674}{14}}$$
$$= \sqrt{162\,548.14} = 403.17$$

Exercise 3.8

Find the standard deviation for the data of Exercise 3.1.

Note: The calculation of $\Sigma(x - \bar{x})^2$ can lead to tedious arithmetic if $\bar{x}$ is not an integer. We may show that $\Sigma(x - \bar{x})^2 = \Sigma x^2 - \frac{(\Sigma x)^2}{n}$ and this simplifies the arithmetic if $\bar{x}$ is not an integer.
For the data of Worked Example no. 3.1 we have already found $\Sigma x = 2\,775$.

$$\Sigma x^2 = 27^2 + 6^2 + 38^2 + 102^2 \dots + 17^2 = 2\,789\,049$$

$$\Sigma x^2 - \frac{(\Sigma x)^2}{n} = 2\,789\,049 - \frac{2\,775^2}{15} = 2\,789\,049 - 513\,375 = 2\,275\,674$$

as before.

Revision Exercises

3.9 The figures below give the overtime earnings of 15 employees of a firm in a week in £'s.
22 18 21 24 19 20 21 17 25 19 23 22 20 23 21
Find (i) mean; (ii) mean deviation; (iii) standard deviation.

3.10 Calculate the arithmetic means of the following sets of data on the weekly wages bills of two small manufacturing firms:

Firm A		*Firm B*	
Wages per week	*No. of employees*	*Wages per week*	*No. of employees*
£40 and under £50	6	£40 and under £50	11
£50 and under £60	12	£50 and under £60	15
£60 and under £70	8	£60 and under £70	16
£70 and under £80	5	£70 and under £80	9
£80 and under £90	14	£80 and under £90	6
£90 and under £100	3	£90 and under £100	2

3.11 Number of children of present marriage according to date of marriage

	Date of Marriage (percentages)	
No. of children	*1949 or earlier*	*1955–1959*
0	3	6
1	20	11
2	31	37
3	19	28
4	13	11
5	6	5
6 or more	8	2

For each distribution calculate the mean and standard deviation.

3.12 **a** Explain the terms:
i mean;
ii standard deviation;
so that the layman can understand.

b Annual mortgage payments:

Amount	*Percentage of mortgages*
Under £80	6
£80 but under £120	6
£120 but under £160	12
£160 but under £200	12
£200 but under £240	11
£240 but under £280	14
£280 but under £320	11
£320 but under £360	8
£360 and over	20

Calculate the mean and standard deviation.

(R.S.A. 1976)

3.13 Journey to work of a sample of 100 London office workers in 1968.

Time of journey (minutes)	*Number of workers*
Under 10	5
10 and less than 20	10

Time of journey (minutes)	*Number of workers*
20 and less than 30	25
30 and less than 40	30
40 and less than 60	17
60 and less than 80	11
80 or more	2
All	100

The above table shows the journey to work times of 100 office workers in a large London firm.

Find the mean and standard deviation of journey time.

3.14 Income of females according to educational qualifications

Annual income	*Qualifications (A or O level GCE)*	*No qualifications*
£ 250 but less than £ 500	8.8	8.0
£ 500 but less than £ 750	15.3	20.6
£ 750 but less than £1 000	23.8	39.2
£1 000 but less than £1 250	25.9	20.0
£1 250 but less than £1 500	18.7	6.6
£1 500 but less than £2 000	5.2	2.2
£2 000 and over	2.3	1.4
	100.0	100.0

Source: *General Household Survey* 1971

Calculate the mean and standard deviation of income for females who have A or O level GCE qualifications.

The mean and standard deviation of income for females who have no qualifications are £913 and £340 respectively.

Compare the two distributions and comment generally.

3.15 a Briefly describe the following measures of dispersion. Discuss their relative advantages and disadvantages. (i) Range, (ii) Quartile deviation; (iii) Mean deviation, (iv) Standard deviation.

b For the following data find each of these measures of dispersion:

9 12 24 18 8 9 18

(I.O.S. November 1975)

3.16 The following table gives the weekly earnings of a random sample of woman workers.

Weekly earnings (£)	*No. of women*
Under 15	1
15 and under 20	4
20 and under 25	28
25 and under 30	42
30 and under 35	33
35 and under 40	18
40 and under 45	13
45 and under 50	9
50 and over	2

i Calculate the mean and the standard deviation of the above distribution.

ii What will happen to the value of the mean and standard deviation if every woman has an increase of (a) £4.40 per week; (b) 10% of previous earnings? (You are not expected to recalculate the actual values in (b).)

(I.O.S. November 1975)

3.17 You have been asked to estimate the population mean value and standard deviation from a sample of 30 values. Your results are:

Mean = 16.4

Standard Deviation = 3.52

However, due to typing errors, two of the figures given to you were incorrect. The incorrect figures had been given as 31 and 21; they should really have been 13 and 12 respectively.

Using the correct figures, calculate new estimates for the population mean value and standard deviation.

(I.O.S. June 1978)

4 Skewness and Coefficient of Variation

Skewness

Distributions of data are either symmetrical as in Figure 4.1 or asymmetrical as in Figures 4.2 and 4.3.

In statistics we call distributions which are asymmetric *skew distributions*. When the longer tail points in the positive direction (Figure 4.2), we say the distribution has *positive skewness*. When the longer tail points in the negative direction (Figure 4.3), we say the distribution has *negative skewness*.

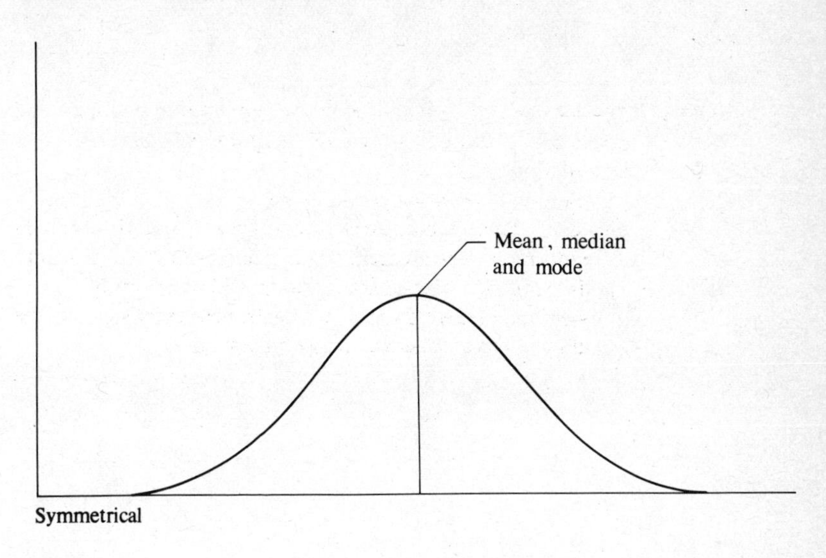

Figure 4.1

In a symmetrical distribution the mean, median and the mode coincide. To measure skewness the difference between the mean and either the mode or the median may be used. Thus a measure of skewness is either

Mean — Mode

or Mean — Median

If we consider Figures 4.4 and 4.5, we see the distributions are of the same shape but differ in scale, hence mean — mode (or mean — median) are different.
When we have distributions of the same shape but different scale we should have similar measures of skewness. We achieve this by dividing by the standard deviation σ to obtain a *coefficient of skewness*. For many distributions the following relationship is approximately true:

$$\text{Mean} - \text{Mode} = 3(\text{Mean} - \text{Median})$$

Hence either $\dfrac{\text{Mean} - \text{Mode}}{\sigma}$ or $\dfrac{3(\text{Mean} - \text{Median})}{\sigma}$

are used as coefficients of skewness. In examination questions you are often asked to find a coefficient of skewness. Which formula to use depends on the information you are given. If you have a choice, choose the formula which enables you to find the measures you can obtain accurately e.g. if you are better at finding the median, use the formula that incorporates the median.

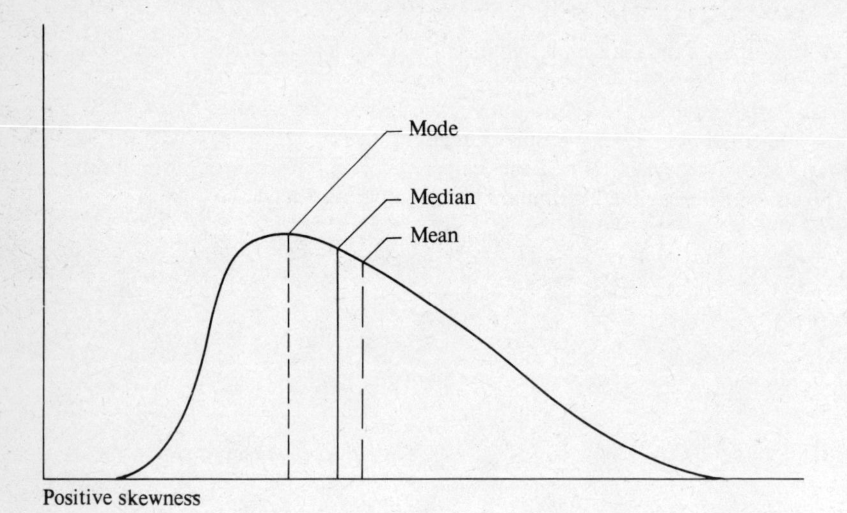

Figure 4.2

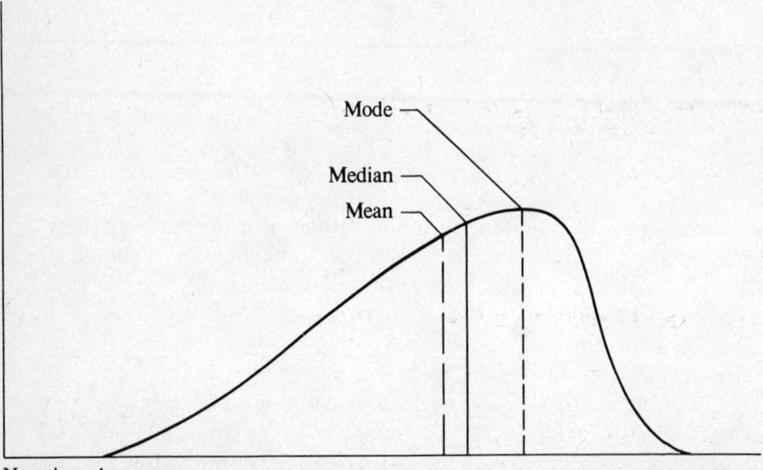

Figure 4.3

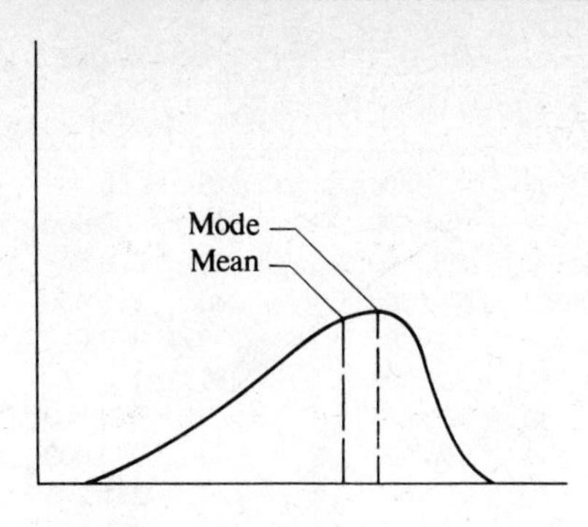

Figure 4.4

Figure 4.5

It should be noted that both the coefficients of skewness mentioned above are independent of the units of the data. For example, it would be possible to compare skewness of income distributions of different countries where the units of currency differ.

Worked Example No. 4.1

Distribution of rents 1972

£ Rents per annum	*Local authority dwellings*	*(Percentages)* *Privately rented unfurnished dwellings*
Under 40	2	17
40 and under 80	15	34
80 and under 120	25	16
120 and under 160	27	9
160 and under 200	18	5
200 and under 240	8	4
240 and under 280	3	4
280 and under 320	2	3
320 and under 360	—	2
360 and under 400	—	3
400 and over	—	2
Total	100	99

Source: *Social Trends* 1974

Calculate mean and standard deviation for both distributions. Find a coefficient of skewness for both distributions.

Range	Central value	(A) *Local authority* f	fx	fx^2	(B) *Privately rented* f	fx	fx^2
0– 40	20	2	40	800	17	340	6 800
40– 80	60	15	900	54 000	34	2 040	122 400
80–120	100	25	2 500	250 000	16	1 600	160 000
120–160	140	27	3 780	529 200	9	1 260	176 400
160–200	180	18	3 240	583 200	5	900	162 000
200–240	220	8	1 760	387 200	4	880	193 600
240–280	260	3	780	202 800	4	1 040	270 400
280–320	300	2	600	180 000	3	900	270 000
320–360	340	—	—	—	2	680	231 200
360–400	380	—	—	—	3	1 140	433 200
400+	420	—	—	—	2	840	352 800
Total		100	13 600	2 187 200	99	11 620	2 378 800

$$\text{(A) Mean} = \frac{\Sigma fx}{\Sigma f} = \frac{13\,600}{100} = £136$$

$$\sigma = \sqrt{\frac{\Sigma fx^2}{\Sigma f} - \left(\frac{\Sigma fx}{\Sigma f}\right)^2}$$

$$= \sqrt{\frac{2\,187\,200}{100} - 136^2}$$

$$= \sqrt{21\,872 - 18\,496}$$

$$= \sqrt{3\,376} = £58.10$$

$$\text{(B) Mean} = \frac{\Sigma fx}{\Sigma f} = \frac{11\,620}{99} = £117.37$$

$$\sigma = \sqrt{\frac{\Sigma fx^2}{\Sigma f} - \left(\frac{\Sigma fx}{\Sigma f}\right)^2}$$

$$= \sqrt{\frac{2\,378\,000}{99} - \left(\frac{11\,620}{99}\right)^2}$$

$$= \sqrt{24\,028.3 - 13\,776.6}$$

$$= \sqrt{10\,251.7} = £101.25$$

To find a measure of skewness we need to find either the mode or the median. In this worked example both measures have been found to illustrate the method.

To find the mode use either graphical method or formulae:

$$Z = L_1 + \left(\frac{f_1 - f_0}{2f_1 - f_0 - f_2}\right)(L_2 - L_1)$$

Local authority

$L_1 = 120$; $L_2 = 160$ (limits of modal group)

f_1 = frequency of modal group = 27

f_0 = frequency of group before modal group = 25

f_2 = frequency of group after modal group = 18

$$Z = 120 + \left(\frac{27 - 25}{2 \times 27 - 25 - 18}\right)(160 - 120) = 120 + \frac{2 \times 40}{11}$$

$$= £127.3$$

Rented privately

$L_1 = 40; L_2 = 80; f_1 = 34; f_0 = 17; f_2 = 16$

$$Z = 40 + \left(\frac{34 - 17}{2 \times 34 - 17 - 16}\right)(80 - 40) = 40 + \frac{17 \times 40}{35}$$

$$= £59.4$$

Skewness Local authority $= \dfrac{\text{Mean} - \text{Mode}}{\sigma} = \dfrac{136 - 127.3}{58.1}$

$$= +0.15$$

Privately rented $= \dfrac{\text{Mean} - \text{Mode}}{\sigma} = \dfrac{117.4 - 59.4}{101.25}$

$$= +0.57$$

To find the median use either a graphical method or a calculation:

	Local authority	*Privately rented*
Less than 40	2	17 ← *M*
Less than 80	17	51
Less than 120	42 ← *M*	67
Less than 160	69	76
Less than 200	87	81
Less than 240	95	85
Less than 280	98	89
Less than 320	100	92
Less than 360		94
Less than 400		97
Less than 440		99

Local authority Median $= 120 + (160 - 120)\left(\dfrac{50 - 42}{69 - 42}\right)$

$$= £131.9$$

Privately rented Median $= 40 + (80 - 40)\left(\dfrac{49.5 - 17}{51 - 17}\right)$

$$= £78.2$$

Skewness Local authority $= \dfrac{3(\text{Mean} - \text{Median})}{\sigma}$

$$= \frac{3(136 - 131.9)}{58.1} = +0.21$$

Privately rented $= \dfrac{3(\text{Mean} - \text{Median})}{\sigma}$

$$= \frac{3(117.4 - 78.2)}{101.25} = +1.16$$

Exercise 4.1

Age of borrower purchasing a house by means of mortgage in 1972

Age (years)	*Percentage of borrowers*
Under 25	21
25 but under 35	45
35 but under 45	21
45 but under 55	10
55 and over	3

Source: *Social Trends*

For the above data calculate median, mean and standard deviation. Hence find a measure of skewness.

Quartile measure of skewness

Skewness may be measured using quartiles and median.

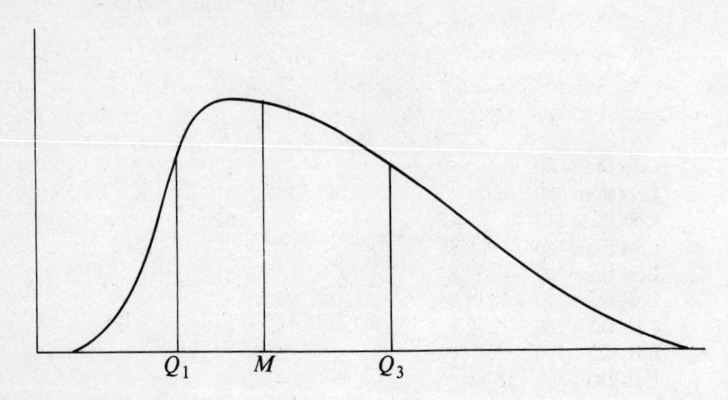

Figure 4.6

If distribution is symmetrical, Q_3 - Median = Median - Q_1
If distribution is asymmetric, (Q_3 - Median) - (Median - Q_1) measures skewness.
To obtain a coefficient of skewness we divide by the inter-quartile range $Q_3 - Q_1$.

$$\text{Thus coefficient of skewness} = \frac{(Q_3 - \text{Median}) - (\text{Median} - Q_1)}{Q_3 - Q_1}$$

$$= \frac{Q_3 + Q_1 - 2 \times \text{Median}}{Q_3 - Q_1}$$

Worked Example No. 4.2

Female employees by highest qualification level and earnings in 1971 (full-time employees)

(*Percentages*)

Annual earnings £	*Degree or equivalent*	*G.C.E. 'A' or 'O' level*
Less than 250	3.8	8.8
250 and less than 500	5.7	15.3
500 and less than 750	7.2	23.8
750 and less than 1 000	12.3	25.9
1 000 and less than 1 250	21.0	18.7
1 250 and less than 1 500	16.7	5.2

Annual earnings £	*Degree or equivalent*	*G.C.F. 'A' or 'O' level*
1 500 and less than 2 000	20.4	1.5
2 000 and less than 2 500	7.2	0.4
2 500 and less than 3 000	3.8	0.4
3 000 and over	1.9	0.0
Total	100.0	100.0

Source: General Household Survey 1971

Calculate the median, lower quartile and upper quartile for both distributions. Compare the two distributions by finding the quartile deviation and coefficient of skewness.

Less than (£)	*Degree*		*'O' or 'A'*	
250	3.8		8.8	
500	9.5		24.1	
750	16.7	← Q_1	47.9	← Q_1
1 000	29.0		73.8	← M
1 250	50.0	← M	92.5	← Q_3
1 500	66.7	← Q_3	97.7	
2 000	87.1		99.2	
2 500	94.3		99.6	
3 000	98.1		100.0	
5 000	100.0			

$$n = 100$$

$$M = \frac{n}{2} = 50$$

$$Q_1 = \frac{n}{4} = 25$$

$$Q_3 = \frac{3n}{4} = 75$$

Degree

$$Q_1 = 750 + (1\,000 - 750)\left(\frac{25 - 16.7}{29 - 16.7}\right) = 750 + \left(\frac{250 \times 8.3}{12.3}\right)$$

$$= 750 + 168.7 = £919$$

$$M = £1\,250$$

$$Q_3 = 1\,500 + (2\,000 - 1\,500)\left(\frac{75 - 66.7}{87.1 - 66.7}\right) = 1\,500 + \left(\frac{500 \times 8.3}{20.4}\right)$$

$$= 1\,500 + 203 = £1\,703$$

O or A

$$Q_1 = 500 + (750 - 500)\left(\frac{25 - 24.1}{47.9 - 24.1}\right) = 500 + \left(\frac{250 \times 0.9}{23.8}\right)$$

$$= 500 + 9 = £509$$

$$M = 750 + (1\,000 - 750)\left(\frac{50 - 47.9}{73.8 - 47.9}\right) = 750 + \left(\frac{250 \times 2.1}{25.9}\right)$$

$$= 750 + 20 = £770$$

$$Q_3 = 1\,000 + (1\,250 - 1\,000)\left(\frac{75 - 73.8}{92.5 - 73.8}\right)$$

$$= 1\,000 + \left(\frac{250 \times 1.2}{18.7}\right)$$

$$= 1\,000 + 16 = £1\,016$$

$$\text{Quartile deviation} = \frac{Q_3 - Q_1}{2}$$

$$\textit{Degree} \quad \frac{Q_3 - Q_1}{2} = \frac{1\,703 - 919}{2} = \frac{784}{2} = £392$$

$$\textit{O or A} \quad \frac{Q_3 - Q_1}{2} = \frac{1\,016 - 509}{2} = \frac{507}{2} = £253$$

$$\text{Skewness} = \frac{Q_3 + Q_1 - 2 \times \text{Median}}{Q_3 - Q_1}$$

$$\textit{Degree} = \frac{1\,703 + 919 - 2 \times 1\,250}{1\,703 - 919} = \frac{122}{784} = +0.156$$

$$\textit{O or A} = \frac{1\,016 + 509 - 2 \times 770}{1\,016 - 509} = \frac{-15}{507} = -0.03$$

Exercise 4.2

Age distribution of the labour force, U.K. 1973

Age in years	*% of total males*	*% of total females*
16 – 19	6.4	1.1
20 – 24	11.0	8.6
25 – 34	22.7	20.2
35 – 44	19.7	25.7
45 – 54	20.4	28.4
55 – 59	8.6	9.3
60 – 64	7.9	4.8
65 +	3.3	1.9
	100.0	100.0

For both distributions calculate:

i the median
ii first and third quartiles and quartile deviation
iii the measure of skewness

Coefficient of Variation

The standard deviation measures variation, but in comparing variation of two distributions with different means it is desirable to consider the standard deviation in relation to the mean. We use a measure of relative variation called the coefficient of variation.

We define the coefficient of variation as $\dfrac{\text{Standard deviation}}{\text{Mean}} = \dfrac{\sigma}{\overline{x}}$

This is usually expressed as a percentage: $\dfrac{\text{Standard deviation} \times 100}{\text{Mean}} = \dfrac{\sigma \times 100}{\overline{x}}$

Note: The coefficient of variation is most useful in comparing two distributions where the data is in different units.

Worked Example No. 4.3

Using the data of Worked Example no. 4.1, compare the two distributions by finding the coefficients of variation.

In Worked Example no. 4.1 we obtained the following results:

	Local authority	*Privately rented*
Mean	136	117.4
Standard deviation	58.1	101.25
Coefficient of variation $= \dfrac{\sigma \times 100}{\overline{x}}$	$\dfrac{58.1 \times 100}{136} = 42.7$	$\dfrac{101.25 \times 100}{117.4} = 86.2$

This shows that privately rented unfurnished dwellings have higher relative variation.

Exercise 4.3

Using the data of Worked Example no. 4.2, relating to earnings by academic qualification, find the mean and standard deviation for both distributions. Compare the distributions by finding the coefficients of variation.

Revision Exercises

4.4 The waiting times of a random sample of 100 customers in a check-out queue at a self-service shop are given below.

Waiting time (minutes)	*Number of customers*
No waiting	5
Under $\frac{1}{2}$	9
$\frac{1}{2}$ but under 1	11
1 but under $1\frac{1}{2}$	18
$1\frac{1}{2}$ but under 2	23
2 but under $2\frac{1}{2}$	16

$2\frac{1}{2}$ but under 3	7
3 but under $3\frac{1}{2}$	5
$3\frac{1}{2}$ but under 4	3
4 minutes and over	3

Calculate the median, mean, standard deviation and coefficient of skewness.

4.5 Data below give distribution of gross weekly earnings for full-time male adults.

Weekly earnings (£)	*% of all full-time male adults*
under 18	1.1
18 - 20	1.1
20 - 25	8.6
25 - 30	14.9
30 - 35	18.2
35 - 40	18.1
40 - 45	14.2
45 - 50	9.7
50 - 60	9.4
60 - 80	4.0
80 and over	0.7
	100%

Source: *New Earnings Survey* 1973

i Calculate arithmetic mean and standard deviation. Comment on your results and on any factors affecting their accuracy. What are the principal uses of the standard deviation?

ii Given that median earnings for the same distribution was £36.60 make any other suitable comments.

(H.N.D. 1976)

4.6 **Age distribution of men and women receiving retirement and old persons' pension in 1976:**

	Percentage	
Age	*Men*	*Women*
60 but under 65		16
65 but under 70	38	25
70 but under 75	32	24
75 but under 80	18	18
80 but under 85	8	11
85 but under 90	3	5
90 and over	1	1

Source: Department of Health and Social Security

i Obtain the median, lower quartile and upper quartile ages for *women*.

ii For *men* the median, lower quartile and upper quartile ages are 71.9, 68.3, and 76.4 respectively. By using suitable measures compare the two distributions.

iii Comment on the data and your results.

(B.A.S.S. (part-time) 1978)

4.7 **Weekly income of households in South-West Region in 1974**

Range of income per week	*Number of households*
Under £10	21
£10 but under £15	57
£15 but under £20	57
£20 but under £30	112
£30 but under £40	133
£40 but under £50	141
£50 but under £60	113
£60 but under £80	150
£80 but under £100	77
£100 and more	75

Source: *Family Expenditure Survey*

a Obtain the lowest decile, lower quartile, upper quartile, quartile deviation, median and measure of skewness for the above distribution.

b For the Greater London area the corresponding values were: lowest decile £15.30, lowest quartile £28.00, upper quartile £76.30, median £52.50. Compare the two distributions and comment on your results.

c Explain briefly why medians and quartiles are often used as measure of 'average' and dispersion in income distribution.

(H.N.D. 1976)

4.8 **Number of rooms per dwelling in the U.K. — 1961 and 1971**

	(Percentages)	
Number of rooms	*1961*	*1971*
1	1	2
2	5	4
3	12	9
4	27	23
5	35	30
6	13	23
7	3	5
8 or more	4	4

Source: Office of Population Censuses and Surveys

a For each distribution calculate the mean, standard deviation and coefficient of variation of number of rooms per dwelling.

b Comment on your results.

(I.O.S. May 1977)

4.9 Length of life of television tubes

Number of hours	*Number of tubes*
Under 2 000	35
2 000 and under 2 400	115
2 400 and under 2 800	142
2 800 and under 3 200	193
3 200 and under 3 600	174
3 600 and under 4 000	157
4 000 and under 5 000	120
5 000 and under 6 000	51
6 000 and over	13
Total	1 000

Source: Information from manufacturer

For the above distribution:

a State what you would use as class-centre for the two open-ended classes, giving reasons;

b Calculate the mean, standard deviation and coefficient of variation;

c Explain the uses of standard deviation and coefficient of variation.

(I.O.S. 1977)

4.10 For the following data calculate the mean, median, mean deviation, standard deviation and a measure of skewness:

16 20 2 9 8 11 15 9 6 4

(R.S.A. 1975)

4.11 Define the following terms and describe the circumstances in which these measures can most usefully be employed: (i) Range; (ii) Standard deviation; (iii) Coefficient of variation.

Calculate these measures for the following data:

8 9 17 11 18 16 14 13 12 9

(I.O.S. Nov. 1973)

5 Regression and Correlation

In many fields, economic, business, sociological, etc., we are interested in finding if there is a relationship between two variables. For example we would agree in general that there is a relationship between savings and income; the larger a person's income, the larger the amount of that income that is saved. How can we find out more about the relationship between savings and income of people living in England? We could choose a random sample of people living in England and ask them to state their savings and income. We could plot the results on graph paper, savings on the Y axis (vertical) and income on the X axis (horizontal). The standard practice is to plot the 'dependent' variable on the Y axis. In our example savings depend on the level of income.

The plot of the results on graph paper is called a *scatter diagram*. The scatter diagram in Figure 5.1 shows that there is some relationship. We know from experience that some people spend most of their income and some save more than others. We would not therefore expect a perfect relationship.

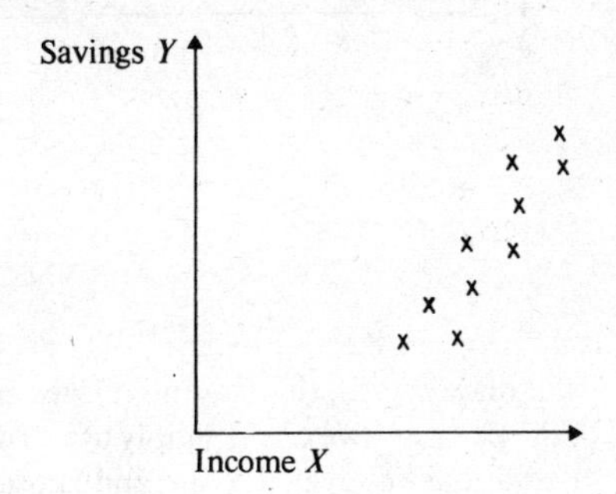

Figure 5.1

The following relationships can be seen in the scatter diagrams overleaf:

i Very good relationship between Y and X, a positive relationship since as X increases Y increases.

ii Some relationship between Y and X, a positive relationship.

iii No relationship between Y and X.

iv Very good negative relationship (or inverse relationship) between Y and X: as X increases Y *decreases*.

v Some negative relationship.

vi A curvilinear relationship (this is too difficult to deal with in an elementary course).

Types of scatter diagram

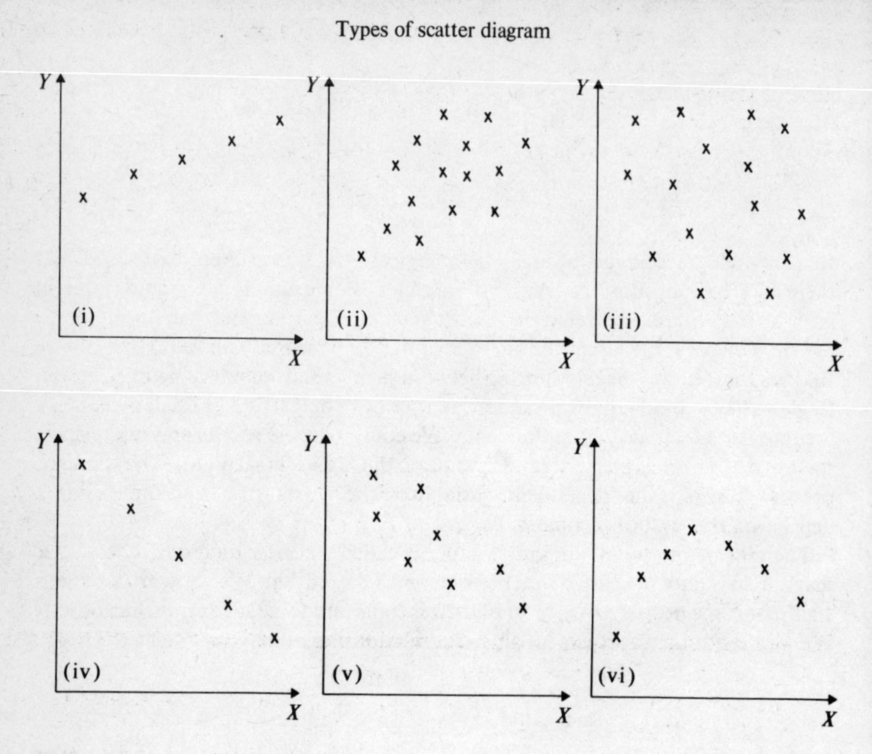

Figure 5.2

Regression line

We can find a relationship between Y and X in terms of the equation of a line which best fits the data. In scatter diagrams (i), (ii), (iv) and (v) we can do this by drawing a line which goes through the points — we could simply use a ruler and draw a line so that the number of observations 'above' the line and those 'below' the line are equal. This is called *fitting a line by eye*.

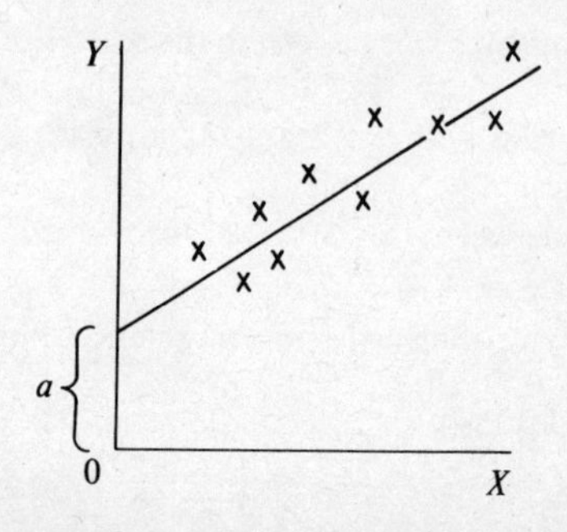

Figure 5.3

It is also possible to fit a line by means of a calculation. In either case the line of best fit is called the *regression line* of Y on X.

The equation of the regression line is usually written in the form

$$Y = a + bX$$

where a is the intercept and is equal to the distance from the origin to the point where the regression line cuts the Y axis, and b is the gradient of the regression line.

We may show that a and b are given by the following formulae (other formulae are sometimes used).

$$b = \frac{n\Sigma(XY) - (\Sigma Y)(\Sigma X)}{n\Sigma X^2 - (\Sigma X)^2} \qquad a = \overline{Y} - b\overline{X}$$

where $\overline{Y} = \dfrac{\Sigma Y}{n}$ and $\overline{X} = \dfrac{\Sigma X}{n}$ and n is the number of pairs of observations.

The regression line $Y = a + bx$ so found is sometimes called the least squares regression line. b is called the *regression coefficient* and may be defined as the change in Y for unit change in X:

e.g. If $b = 2$, then the change in Y is 2 when X increases by 1.
If the line slopes upwards, b is positive.
If the line slopes downwards, b is negative.

Important note

In examination questions you are often not given which set of data is 'Y' and which is 'X'. It is vital that you choose Y and X correctly.

a If you are asked to find the regression of 'A' on 'B' then the 'A' data must be the 'Y' and 'B' the 'X'.

b If you are asked to predict a future value or to estimate the value of the variable when the other variable takes a given value, then the predictor variable or the estimated variable must be the 'Y' data.

c In many cases the 'dependent' variable is obvious. The dependent variable is the Y data.

Worked Example No. 5.1

Month	*Mean temperature* (°C)	*Energy consumption* (*Millions of tons of coal equivalent*)
January	7	29
February	5	30
March	5	34
April	9	27
May	10	23
June	15	25
July	18	19
August	19	18
September	14	17
October	11	26
November	7	28
December	5	35

Source: *Monthly Digest of Statistics*

a Calculate the equation of the least squares regression which will enable you to estimate energy consumption from a given temperature.

b Estimate the energy consumption when the temperature is 12°C.

It is essential to choose Y and X correctly. Here we are asked to estimate energy consumption, so energy consumption is the Y data. Also energy consumption depends on the temperature. Energy consumption is the dependent variable and thus the Y variable.

X	Y	XY	X^2
7	29	203	49
5	30	150	25
5	34	170	25
9	27	243	81
10	23	230	100
15	25	375	225
18	19	342	324
19	18	342	361
14	17	238	196
11	26	286	121
7	28	196	49
5	35	175	25
125	311	2 950	1 581

$$b = \frac{n\Sigma YX - (\Sigma Y)(\Sigma X)}{n\Sigma X^2 - (\Sigma X)^2} = \frac{12 \times 2\,950 - 125 \times 311}{12 \times 1\,581 - (125)^2}$$

$$= \frac{35\,400 - 38\,875}{18\,972 - 15\,625} = -\frac{3\,475}{3\,347} = -1.038$$

$$\overline{Y} = \frac{\Sigma Y}{n} = \frac{311}{12} = 25.92$$

$$\overline{X} = \frac{\Sigma X}{n} = \frac{125}{12} = 10.42$$

$$a = \overline{Y} - b\overline{X} = 25.92 - (-1.038) \times 10.42 = 25.92 + 10.82$$

$$= 36.74$$

$$Y = a + bX \qquad \therefore\ Y = 36.74 - 1.038X$$

This is a downward-sloping regression line; as X increases Y decreases. In this case we would expect energy consumption to fall when the temperature rises. The regression coefficient (−1.038) shows that for each increase of one degree centigrade in temperature monthly energy consumption falls by 1.038 million tons of coal equivalent.

To find the energy consumption when the temperature is 12°C, put $X = 12$ in the regression equation.

$$Y = 36.74 - 1.038 \times 12 = 36.74 - 12.46 = 24.28$$

Thus estimate of energy consumption is 24 million tons.

Exercise 5.1

Find the regression line of Y on X for the following data:

X	5	7	9	11	13	15
Y	1.7	2.4	2.8	3.4	3.7	4.4

Use of an assumed mean

In some situations the arithmetic can be very heavy and 'overflow' can occur on some calculators. The following example shows how to deal with large numbers.

Worked Example No. 5.2

Expenditure on education between 1967 and 1976 at 1970 prices

Year	*Expenditure* (£ *million*)
1967	1 484
1968	1 558
1969	1 599
1970	1 685
1971	1 763
1972	1 903
1973	2 023
1974	2 083
1975	2 158
1976	2 277

Source: *National Income and Expenditure*

Calculate the regression equation which will enable you to forecast future expenditure on education. Use your regression equation to forecast expenditure in 1980.

Expenditure is Y data. To cut down the arithmetic, let 1967 be Year 1 and give Y data to 3 significant figures.

X	Y	X^2	XY
1	148	1	148
2	156	4	312
3	160	9	480
4	169	16	676
5	176	25	880
6	190	36	1 140
7	202	49	1 414
8	208	64	1 664
9	216	81	1 944
10	228	100	2 280
55	1 853	385	10 938

$$b = \frac{n\Sigma YX - (\Sigma Y)(\Sigma X)}{n\Sigma X^2 - (\Sigma X)^2} = \frac{10 \times 10\,938 - 1\,853 \times 55}{10 \times 385 - 55^2}$$

$$= \frac{109\,380 - 101\,915}{3\,850 - 3\,025} = \frac{7\,465}{825} = 9.048$$

$$\overline{Y} = \frac{\Sigma Y}{n} = \frac{1\,853}{10} = 185.3$$

$$\overline{X} = \frac{\Sigma X}{n} = \frac{55}{10} = 5.5$$

$$a = \overline{Y} - b\overline{X} = 185.3 - 9.048 \times 5.5 = 185.3 - 49.8 = 135.5$$

$$\therefore \quad Y = 135.5 + 9.048X$$

1980 (with 1967 = 1) is year 14. To find the forecast for 1980 put $X = 14$.

$$Y = 135.5 + 9.048 \times 14 = 135.5 + 126.7 = 262$$

Thus forecast expenditure on education in 1980 is £2 620 million.

Exercise 5.2

Average purchase price of houses in Greater London

Year	*Average purchase price* (*£ thousand*)
1969	6.2
1970	6.9
1971	7.1
1972	11.1
1973	14.4
1974	14.9
1975	14.9

Source: *Social Trends*

a Plot a scatter diagram for the above data.

b Calculate the least squares linear regression line of the average purchase price on year and draw it on your scatter diagram.

c **i** Forecast the average price of houses in Greater London in 1977 using your regression line.

ii Is this a realistic forecast? Discuss briefly the statistical problems of producing short-term forecasts of average purchase price of houses in Greater London.

(B.A.B.S. 1977)

Correlation coefficient

For scatter diagrams (i) and (ii) of figure 5.2 we can calculate a regression line $Y = a + bX$. Obviously the regression line will fit diagram (i) better than diagram (ii). We want to have some measure of how good a fit a line has to data. The correlation coefficient provides us with this measure.

The product moment correlation coefficient r is found from the formula (other formulae are sometimes used)

$$r = \frac{n\Sigma(YX) - (\Sigma Y)(\Sigma X)}{\sqrt{(n\Sigma Y^2 - (\Sigma Y)^2)(n\Sigma X^2 - (\Sigma X)^2)}}$$

r *must* be between $+1$ and -1. If you obtain an answer numerically larger than 1 you have made an arithmetic error.

Interpretation of the value of r

If $r = +1$ (scatter diagram Figure 5.2 (i)) there is a perfect direct relationship between Y and X.

If $r = -1$ (scatter diagram Figure 5.2 (iv)) there is a perfect inverse (or negative) relationship between Y and X.

If $r = 0$ (scatter diagram Figure 5.2 (iii)), there is no relationship between Y and X.

If r = a value between 0 and 1 (scatter diagram Figure 5.2 (ii)) there is some positive relationship between Y and X. The nearer r is to 1, the stronger the relationship.

If r = a value between 0 and -1 (scatter diagram Figure 5.2 (v)), there is some inverse relationship between Y and X. The nearer r is to -1, the stronger the inverse relationship.

NB You can work from an assumed mean for Y and/or X. If Y and X are large, 'overflow' problems can occur on calculators.

A correlation coefficient examines whether data shows a linear relationship (i.e. lies on a straight line). A high value of r (near to $+1$ or -1) shows that the data lies close to a straight line. A low value of r, particularly when n is small, is not very useful; the data could be as in Figure 5.4 or as in Figure 5.5.

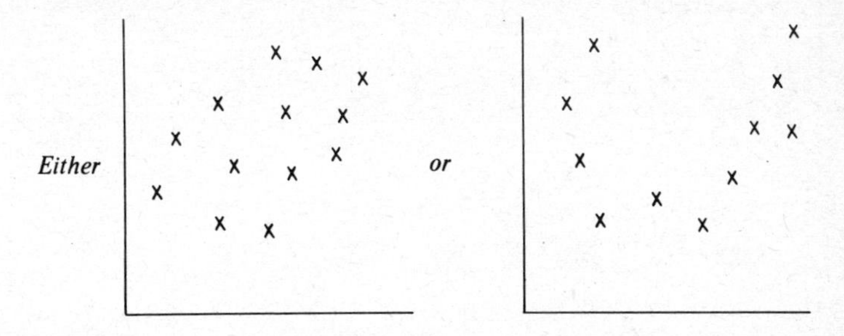

Figure 5.4 **Figure 5.5**

In both cases we would obtain a low value of r. The interpretation of r is usually considerably improved if a scatter diagram is drawn.

On the other hand, if r is high, this shows that Y and X are related, but *it does not* prove that they are *causally* related. This may be summed up by the phrase 'correlation does not prove causation'.

Worked Example No. 5.3

Draw a scatter diagram and calculate the product moment correlation coefficient for the pairs (XY) given below:

X:	1	3	4	7	10	12	13	14
Y:	9	7	3	2	5	7	10	10

X	Y	X^2	Y^2	XY
1	9	1	81	9
3	7	9	49	21
4	3	16	9	12
7	2	49	4	14
10	5	100	25	50
12	7	144	49	84
13	10	169	100	130
14	10	196	100	140
64	53	684	417	460

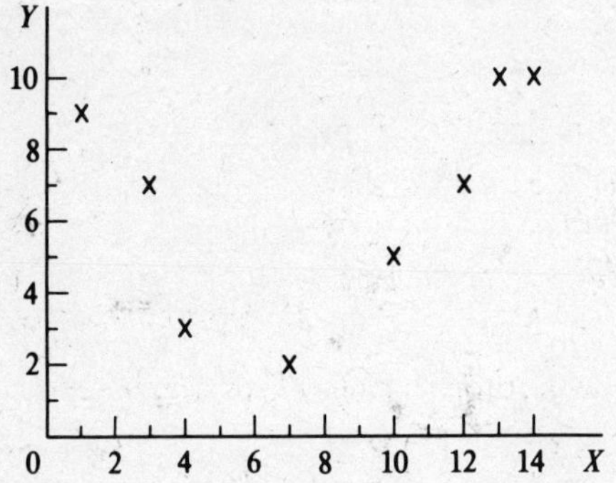

Figure 5.6

$$
\begin{aligned}
r &= \frac{n\Sigma(YX) - (\Sigma Y)(\Sigma X)}{\sqrt{(n\Sigma Y^2 - (\Sigma Y)^2)(n\Sigma X^2 - (\Sigma X)^2)}} \\
&= \frac{8 \times 460 - 53 \times 64}{\sqrt{(8 \times 417 - 53^2)(8 \times 684 - 64^2)}} \\
&= \frac{3\ 680 - 3\ 392}{\sqrt{(3\ 336 - 2\ 809)(5\ 472 - 4\ 096)}} \\
&= \frac{288}{\sqrt{527 \times 1\ 376}} \\
&= \frac{288}{\sqrt{725\ 152}} \\
&= \frac{288}{851.6} = 0.338 = 0.34
\end{aligned}
$$

Important note

If we have large numbers for Y and X, the number in the square root can sometimes be too big for some calculators to handle. We proceed as follows using the example above.

$$r = \frac{288}{\sqrt{527 \times 1\,376}} = \frac{288}{\sqrt{527} \times \sqrt{1\,376}} = \frac{288}{22.96 \times 37.09}$$
$$= \frac{288}{851.6} = 0.34$$

Exercise 5.3

1976	*Mean temperature* (°C)	*Beer production* (*million barrels*)
January	6	2.5
February	5	2.4
March	5	3.3
April	8	3.3
May	12	3.5
June	17	3.7
July	19	3.9
August	18	3.6
September	14	3.4
October	11	3.1

Source: *Monthly Digest of Statistics*

Calculate the product moment correlation coefficient.

Rank Correlation Coefficient

In some situations data is given ranked. In this situation the calculation of the correlation coefficient is simplified if the rank correlation coefficient is used.

The rank correlation coefficient is given by

$$r = 1 - \frac{6\Sigma d^2}{n^3 - n}$$

where d is the difference in ranks and n is the size of the sample.

The use of the rank correlation coefficient is best illustrated by means of examples.

Worked Example No. 5.4

In a drama competition 10 plays were ranked by two adjudicators as follows:

Play	A	B	C	D	E	F	G	H	J	K
X	5	2	6	8	1	7	4	9	3	10
Y	1	7	6	10	4	5	3	8	2	9

Calculate the rank correlation coefficient.

$d = 4 \quad 5 \quad 0 \quad 2 \quad 3 \quad 2 \quad 1 \quad 1 \quad 1 \quad 1$

$\Sigma d^2 = 16 + 25 + 0 + 4 + 9 + 4 + 1 + 1 + 1 + 1 = 62$

$n = 10$

$$r = 1 - \frac{6 \times 62}{10^3 - 10} = 1 - \frac{372}{990} = 0.624$$

Exercise 5.4

Define a coefficient of *rank* correlation and discuss its uses. Calculate a rank correlation coefficient for the following data.

		Test I	*Test II*
Student	A	7	8
	B	3	4
	C	5	6
	D	8	5
	E	2	1
	F	10	10
	G	1	3
	H	9	9
	I	6	6
	J	3	2

(H.N.D. 1977)

When actual measurements of X and Y are given for some data it is possible to calculate either a product moment or a rank correlation coefficient but it is necessary to rank the data in order to calculate a rank correlation coefficient.

Worked Example No. 5.5

The following data give floor space per employee and turnover for 8 firms.

Firm	A	B	C	D	E	F	G	H
Floor space (m^2)	60	55	40	30	70	60	65	50
Turnover (£1 000)	22	20	17	18	25	23	21	19

The first step is to put data into rank order.

Floor space	$3\frac{1}{2}$*	5	7	8	1	$3\frac{1}{2}$*	2	6
Turnover	3	5	8	7	1	2	4	6
d	$\frac{1}{2}$	0	1	1	0	$1\frac{1}{2}$	2	0

$\Sigma d^2 \quad \frac{1}{4} + 0 + 1 + 1 + 0 + 2\frac{1}{4} + 4 + 0 = 8\frac{1}{2}$

$$r = 1 - \frac{6\Sigma d^2}{n^3 - n} = 1 - \frac{6 \times 8\frac{1}{2}}{8^3 - 8}$$

$$= 1 - \frac{51}{512 - 8} = 1 - \frac{51}{504} = 1 - 0.101 = 0.899$$

* In the case of ties the simplest method is to give ties half the sum of the ranks. In this example 60 appears twice and would have 3rd and 4th rank. Both are given the rank $\frac{3 + 4}{2} = 3\frac{1}{2}$.

N.B. As an exercise, show that the product moment correlation coefficient for the actual data is 0.870. It should be noted that the two results are nearly equal. If you are asked to find both a product moment correlation coefficient and a rank correlation coefficient and your answers are considerably different, you have probably made a mistake and you should check your working.

Exercise 5.5

The table below shows the number of crimes committed in a sample of local authority areas during a certain period:

Population of local authority (10 000s)	*Number of crimes committed*
5	12
7	9
8	14
11	20
13	13
15	14
16	22
19	21
22	29
23	28

For the data above, calculate:

a The product moment correlation coefficient; and

b A rank correlation coefficient.

Indicate how you would interpret these coefficients and which you would regard as the more reliable statistic for the purpose of interpreting these data.

(B.Sc.(Soc.) 1966 Ext.)

Important note

The usual examination questions on correlation and regression have now been covered. In some examinations questions are set involving (i) the coefficient of determination or (ii) the significance of the product moment correlation coefficient. In the case of the significance of the correlation coefficient it is essential to read Chapter 10 first. Exercises on these topics are found at the end of the chapter.

Coefficient of determination (r^2)

In correlation and regression analyses we are trying to find how much of the change in the values of the Y data can be 'explained' by the regression line of Y on X. If r is low, as in diagram (ii) of Figure 5.2, a regression line is not a very good fit to the data. If r is high, as in diagram (i), a regression line is a good fit to the data.

Suppose we are investigating the factors that account for variation in sales of a product in various areas. Let the sales be the Y data; then from our sample data the total variation in sales is $\Sigma(Y - \overline{Y})^2$. If for each area we have X as the amount spent on advertising, we could then plot a scatter diagram of the data. We would expect that as X increases Y increases, i.e. a positive correlation. We would not, however, expect perfect correlation since, if the product is income-elastic, we would find above-average sales in wealthy areas and below-average sales in poor areas.

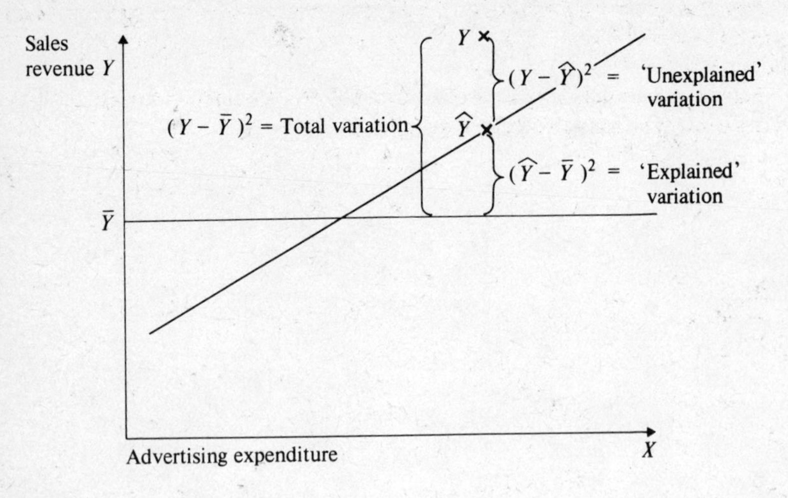

Figure 5.7

The regression line 'explains' some of the variation in Y, i.e. as X increases Y increases. In the diagram (Figure 5.7) the 'explained' variation for one observation is $(\hat{Y} - \overline{Y})^2$. For all observations the 'explained' variation is $\Sigma(\hat{Y} - \overline{Y})^2$.

The vertical distance of an observation from the regression line is $Y - \hat{Y}$ and the 'unexplained' variation is $(Y - \hat{Y})^2$. For all observations, the 'unexplained' variation is $\Sigma(Y - \hat{Y})^2$.

It is possible to show that

$$r^2 = \frac{\Sigma(\hat{Y} - \overline{Y})^2}{\Sigma(Y - \overline{Y})^2} = \frac{\text{Variation of } Y \text{ 'explained' by regression line}}{\text{Total variation of } Y}$$

i.e. the fraction of the total variation of Y which is 'explained' by the regression line and therefore by the variation of X is r^2. r^2 is called the *coefficient of determination*.

It is evident that small values of r, even if significant, need to be interpreted with caution. If $r = 0.7$, then $r^2 = 0.49$ and thus 49% of the variation of Y is 'explained' by the regression line, i.e. by the variation of X. Thus if for sales revenue and advertising expenditure the correlation coefficient were 0.7, 49% of the variation in sales revenue would be accounted for by advertising expenditure and 51% would be due to other factors.

If $r = 0.2$ (and from the table on page 90 this value of r is significant at 5% if we have about 100 observations), $r^2 = 0.04$. This means that 4% of the variation of Y is 'explained' by the variation of X and 96% of the variation of Y is 'unexplained'. Thus at a value of $r = 0.2$, with just 4% of the variation of Y 'explained' by variation in X, Y is almost independent of X.

Significance of the product moment correlation coefficient

When we obtain the value of the correlation coefficient from our *sample* data, we want to find out whether there is likely to be a relationship between the two variables in the *population*.

Suppose we have a random sample of 10 pairs of observations and we obtain a correlation coefficient of 0.7. Is there any evidence of a relationship in the population?

For our sample data with 10 observations and $r = 0.7$ the scatter diagram for our sample might be something like Figure 5.8.

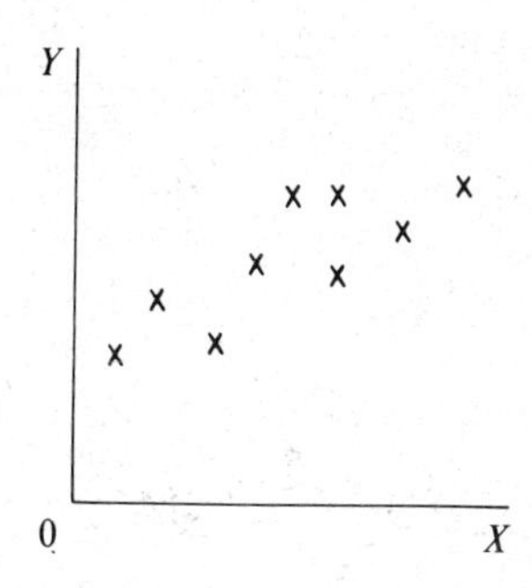

Figure 5.8

The question we want to resolve is whether this sample is likely to be drawn from population A or population B. See Figures 5.9 and 5.10 overleaf.

It seems likely that the sample is drawn from population B but it is possible that it may have been drawn from population A.

It is possible to work out the probability that a sample of 10 observations with $r = 0.7$ is drawn from population A. This is rather advanced but special tables have been worked out.

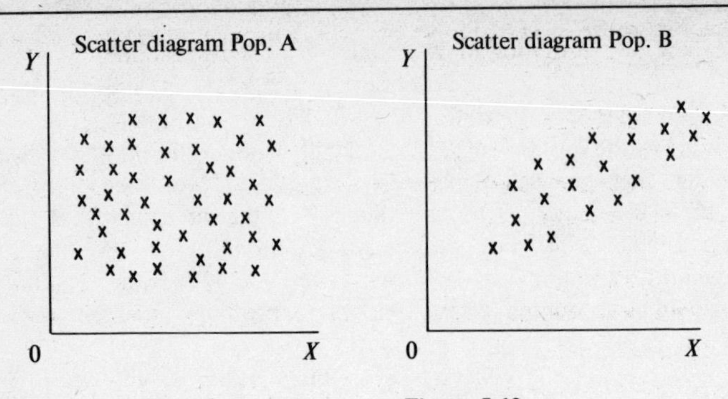

Figure 5.9 **Figure 5.10**

Population A clearly has zero correlation
Population B has a definite positive correlation.

n	8	10	12	14	16	18	20	22	32	42	52	102
5%	.707	.632	.576	.532	.497	.468	.444	.423	.349	.304	.273	.195
1%	.834	.765	.707	.661	.623	.590	.561	.537	.449	.393	.354	.254

Suppose that the population correlation coefficient is ρ. The procedure we adopt is to test the null hypothesis $\rho = 0$ against the alternative hypothesis $\rho \neq 0$. If we have 10 observations the table gives:

$r = 0.632$ 5%
$r = 0.765$ 1%

This means that if $r = 0.632$ there is a one in twenty chance that the sample has been drawn from a population with no relationship between Y and X (e.g. Figure 5.9), i.e. $\rho = 0$, and nineteen chances in twenty that the sample has been drawn from a population with some relationship between Y and X (e.g. Figure 5.10), i.e. $\rho \neq 0$. The usual convention is to accept the null hypothesis $\rho = 0$ if the value of r obtained from a sample is *less* than the 5% point and reject the null hypothesis if the value of r is numerically *greater* than the 5% point and thus conclude that the alternative hypothesis is true, i.e. $\rho \neq 0$. The larger the value of r in relation to the 5% and 1% points, the more confident we are in rejecting the null hypothesis.

Consider again the earlier example with $n = 10$, $r = 0.7$. This value of r is greater than the 5% point (0.632) but less than the 1% point (0.765). We would say that the correlation coefficient r is significant at 5%; hence there is some evidence that $\rho \neq 0$, i.e. some evidence of a relationship between Y and X in the population.

Revision Exercises

5.6 Calculate the regression coefficient and the regression line of Y on X for the following pairs of Y and X.

Y	9	7	4	5	3	2
X	2	4	6	8	10	12

5.7 Find the correlation coefficient for the data of Exercise 5.6.

5.8 **Total H.P. debt outstanding to finance houses and retailers (£000m) (seasonally unadjusted)**

1973	*Mar.*	*Apr.*	*May*	*Jun.*	*Jul.*	*Aug.*	*Sep.*	*Oct.*	*Nov.*	*Dec.*
	2.17	2.21	2.23	2.26	2.28	2.33	2.36	2.40	2.45	2.48
1974	*Jan.*									
	2.43									

Source: *Monthly Digest of Statistics*

Required:

1 Calculate the least squares trend equation for the above series.

2 Use your equation to forecast the H.P. debt for February 1974.

3 Given that the actual February figure was £2.398, comment on the accuracy or otherwise of your forecast.

4 Graph the series from March 1973 to January 1974 together with your trend line.

(A.F.C. 1975)

5.9 **a** Draw scatter diagrams which show:

i Zero correlation;
ii Strong positive correlation;
iii Weak negative correlation.

b Find the rank correlation coefficient for the following data:

Drama competition order of merit

Play:	A	B	C	D	E	F	G	H	I	J
Adjudicator 1	3	7	10	2	9	1	5	4	6	8
Adjudicator 2	1	10	9	4	8	3	2	7	5	6

5.10 Sales trainees take an aptitude test at the end of a course on salesmanship. Following there are the scores (X) obtained by each trainee and his corresponding sales (Y) in thousands of £s for the year following the course.

Find the linear regression equation of Y on X and explain its meaning.

Use the equation to estimate expected sales for a trainee who obtains a score of 70, commenting on any factors affecting the accuracy of your estimate.

X:	18	26	28	34	36	42	48	52	54	60
Y:	54	64	54	62	68	70	76	66	76	74

(H.N.D. 1977)

5.11 Examine the degree of similarity of regional unemployment patterns of males and females by calculating the product-moment correlation coefficient and the rank correlation coefficient for the following data and comment on your results.

	Males	*Females*
	(Percentage unemployed)	
South East	1.8	0.4
Greater London	1.6	0.4
East Anglia	2.2	0.6
South West	3.0	1.0
West Midlands	2.2	0.7
East Midlands	2.4	0.6
Yorkshire	3.2	0.8
North West	4.0	1.0
North	5.3	1.7
Wales	4.0	1.6

(H.N.D. 1976)

5.12 **Gross weekly earnings of non-manual workers by age in 1974**

Age (years)	*Median weekly earnings* (£)
18	15.50
20	23.20
22	34.00
27	44.90
35	53.10
45	55.00
55	51.20

Source: *New Earnings Survey*

a Calculate the least squares regression line of median weekly earnings on age, giving your results in the form $Y = a + bX$.

b *Without recalculation*: (i) Find the values of a and b if all non-manual workers had an increase of £6.00 per week; and (ii) Find the value of b if all non-manual workers had an increase of 6% of their previous earnings.

(I.O.S. May 1977)

5.13 Females as percentages of total economically active and divorce rates per 100,000 population:

	Females as percentage of total economically active	*Divorce rates per 100,000 population*
United Kingdom	37	113
Belgium	33	68
Denmark	40	193
Germany (Federal Republic)	37	124
Luxembourg	26	64
Netherlands	26	74
Japan	39	93

Source: *Social Trends*

i Plot the data on a scatter diagram.
ii Calculate the product moment correlation coefficient between females as percentage of total economically active and divorce rates.
iii Comment on your results.
iv Briefly outline the problems found in correlation and regression analysis.

(B.A.S.S. (P/T) 1978)

5.14 Housing subsidies and tax relief on mortgages

Year	*Average housing subsidy contribution per dwelling* (£)	*Average tax relief to mortgagors* (£)
1967	39	42
1968	42	44
1969	49	50
1970	53	61
1971	61	79
1972	57	70
1973	68	98
1974	135	133

Source: *Social Trends*

a Calculate the product moment correlation coefficient between average housing subsidy and average tax relief to mortgagors. Comment on your result.
b Explain the uses of the coefficient of correlation and the coefficient of determination.

(B.A.S.S. (P/T) 1977)

6 Index Numbers

An index number measures changes in the price, quantity or value of a group of articles or commodities. In the case of a price index, the current prices of a group of articles are compared with the corresponding prices of that group some time in the past. This time is called the *base date*. In the case of the index numbers of wholesale prices, the base date is 1975 = 100.

In the case of the general index of retail prices we find the current prices of the 'basket' of goods and services purchased by a notional household and compare with the prices of the same 'basket' of goods purchased at the base date, which at the time of writing is January 1974. To construct a price index number by this method we need to know the current and the base year prices and the quantities involved in the 'basket' of goods and services.

In the case of imports and exports we are interested not only in price change but in the change of *volume* so in this case we have also volume index numbers.

Unweighted index numbers

Worked Example No. 6.1

Suppose that a housewife purchases food for lunch — rumpsteak, potatoes and Brussels sprouts — and we wish to find out how much this cost in January 1978 and compare it with the cost in January 1972.

	Average price of one pound (p)	
	January 1972	*January 1978*
Rumpsteak	67.2	158.9
Potatoes	2.2	4.7
Brussels Sprouts	4.9	10.8

Data obtained from *Department of Employment Gazette* for February 1972 and 1978

The average cost in January 1978 was 158.9 + 4.7 + 10.8 = 174.4
The average cost in January 1972 was 67.2 + 2.2 + 4.9 = 74.3
We compare the cost in 1978 with the corresponding cost in 1972 by means of the ratio

$$\frac{174.4}{74.3} = 2.347$$

Thus the cost has increased in price by a factor of 2.347 since the 'base date' of January 1972.

It is the convention to express the base date in terms of 100 and in consequence the price index in January 1978 with January 1972 = 100 is 2.347 × 100 = 234.7.
In mathematical terms:
If P_n represents the current year's price
and P_o represents the base year's price
an unweighted index number is $\dfrac{\Sigma P_n}{\Sigma P_o} \times 100$

Exercise 6.1

Average price of one pound (new pence)

	November 1967	March 1970
Rumpsteak	$43\frac{1}{2}$	54
Potatoes	$1\frac{1}{2}$	2
Tomatoes	11	$16\frac{1}{2}$
Cauliflower	7	10

From the above data find an unweighted price index number for March 1970 with November 1967 = 100.

Weighted index number

In Worked Example No. 6.1 it is assumed that the housewife purchased one pound of each item. In practice she will have purchased varying quantities, e.g. $\frac{3}{4}$ lb steak, 2 lb potatoes and 1 lb Brussels sprouts in January 1972.

	Quantity Purchased (lb)	*Average price of one pound (p)* *January 1972*	*January 1978*
Rumpsteak	$\frac{3}{4}$	67.2	158.9
Potatoes	2	2.2	4.7
Brussels sprouts	1	4.9	10.8

Cost in January 1972 = $(\frac{3}{4} \times 67.2) + (2 \times 2.2) + (1 \times 4.9) = 59.7$
Cost in January 1978 = $(\frac{3}{4} \times 158.9) + (2 \times 4.7) + (1 \times 10.8) = 139.4$
The ratio of the cost of this basket of goods in January 1978 compared to the cost in January 1972 is

$$\frac{139.4}{59.7} = 2.335$$

and using the convention that January 1972 = 100, the price index in January 1978 was 2.335 × 100 = 233.5.

In mathematical terms:
If q_n represents the current year quantities
and q_o represents the base year quantities

a *base* year weighted price index number is $\frac{\Sigma q_o p_n}{\Sigma q_o p_o} \times 100$

It is called a *base year weighted index number* because base year quantities are taken as the constant 'basket' of goods. This index number is called a *Laspeyres index.*

Most index numbers in current use in economic statistics are of the Laspeyres form. However, it is possible to use the *current year quantities* as the basket of goods and we have the following:

A current year weighted price index number is $\frac{\Sigma q_n p_n}{\Sigma q_n p_o} \times 100$

This index number is called a *Paasche index.*

Worked Example No. 6.2

a Discuss the relative advantages and disadvantages of Laspeyres and Paasche indices.

b For the following data find (i) Laspeyres price index; (ii) Paasche price index for January 1973, with January 1968 = 100:

	January 1968		*January 1973*	
Commodity	*Average price (new pence) per unit*	*Average quantity purchased*	*Average price (new pence) per unit*	*Average quantity purchased*
Eggs	2	5	2.0	4
Potatoes	2	3	2.5	3
Bread	8	2	10.5	2

a *Laspeyres index number*

Advantages:

1. Quantities have to be found for the base year only.
2. Denominator is constant for all years, hence comparisons can be made between different years as well as with the base year.

Disadvantages:

1. Base year quantities used as weights may not represent current consumption patterns.
2. In general a Laspeyres index overstates the rise in prices. If the price of a particular commodity has an above-average increase in price, then consumers will tend to buy less of this commodity than at base year. Thus the numerator will tend to be larger than it is in practice.

Paasche index number

Advantages:

As current year quantities are used as weights, the index reflects current expenditure patterns.

Disadvantages:

1. Quantities have to be found for each year.
2. The denominator changes each year; thus each year can be compared with the base year only and not other years.
3. In general a Paasche index number understates the actual rise in prices.

b i Laspeyres price index is $\dfrac{\Sigma p_n q_o}{\Sigma p_o q_o} \times 100$

ii Paasche price index is $\dfrac{\Sigma p_n q_n}{\Sigma p_o q_n} \times 100$

It is necessary to calculate $\Sigma p_n q_o$, $\Sigma p_o q_o$, $\Sigma p_n q_n$, $\Sigma p_o q_n$.

p_o	q_o	p_n	q_n	$p_n \times q_o$	$p_o \times q_o$	$p_n \times q_n$	$p_o \times q_n$
2	5	2	4	10	10	8	8
2	3	2.5	3	7.5	6	7.5	6
8	2	10.5	2	21	16	21	16
				38.5	32	36.5	30

Laspeyres price index is $\dfrac{\Sigma p_n q_o}{\Sigma p_o q_o} \times 100 = \dfrac{38.5}{32} \times 100 = 120.3$

Paasche price index is $\dfrac{\Sigma p_n q_n}{\Sigma p_o q_n} \times 100 = \dfrac{36.5}{30} \times 100 = 121.7$

Exercise 6.2

For the following data calculate Paasche and Laspeyres price index numbers.

	1969		*1974*	
	Price per unit (new pence)	*Quantity (no. of units)*	*Price per unit (new pence)*	*Quantity (no. of units)*
Beef	35	3	80	2
Potatoes	2	2	4.5	3
Peas	5	$1\frac{1}{2}$	9	1

(I.O.S. Preliminary May 1975)

Quantity index number

In some situations it is necessary to find a quantity index

A Laspeyres quantity index is $\dfrac{\Sigma q_n p_o}{\Sigma q_o p_o} \times 100$

A Paasche quantity index is $\dfrac{\Sigma q_n p_n}{\Sigma q_o p_n} \times 100$

Worked Example No. 6.3

Find **i** Laspeyres quantity index, **ii** Paasche quantity index. For the data of Worked Example No. 6.2.

i Laspeyres quantity index is $\dfrac{\Sigma q_n p_o}{\Sigma q_o p_o} \times 100$

ii Paasche quantity index is $\dfrac{\Sigma q_n p_n}{\Sigma q_o p_n} \times 100$

We have already found $\Sigma q_n p_o$ etc. in Worked Example No. 6.2.

Laspeyres quantity index is $\dfrac{\Sigma q_n p_o}{\Sigma q_o p_o} \times 100 = \dfrac{30}{32} \times 100 = 93.8$

Paasche quantity index is $\dfrac{\Sigma q_n p_n}{\Sigma q_o p_n} \times 100 = \dfrac{36.5}{38.5} \times 100 = 94.8$

Exercise 6.3

For the data of Exercise 6.2 find Laspeyres and Paasche quantity index numbers.

Important note

In Worked Example No. 6.2 we were given prices and quantities. In examination questions you may sometimes be given price and *expenditure* for each commodity. It is a common fault for candidates not to read the question carefully and think that expenditure means quantity. Expenditure is equal to price × quantity; hence to find quantity it is necessary to divide expenditure by price.

Worked Example No. 6.4

In a factory the wages and the wage bill for manual workers in 1975 and 1978 are given below:

Workers	*1975*		*1978*	
	Weekly wages (£)	*Total weekly wages bill* (£)	*Weekly wages* (£)	*Total weekly wages bill* (£)
Skilled manual	60	2 400	90	3 150
Semi-skilled manual	50	3 000	80	5 600
Unskilled manual	40	2 000	60	2 400

Find (i) a base weighted index of wages (ii) a current weighted index of wages.

The first step is to find the number of employees in each category. In 1975 the number of skilled manual workers was 2 400 ÷ 60 = 40. We do similar calculations for the other categories.

	1975		1978	
	p_o	q_o	p_n	q_n
Skilled manual	60	40	90	35
Semi-skilled manual	50	60	80	70
Unskilled manual	40	50	60	40

We now proceed as in Worked Example No. 6.2.

p_o	q_o	p_n	q_n	p_oq_o	p_oq_n	p_nq_o	p_nq_n
60	40	90	35	2 400	2 100	3 600	3 150
50	60	80	70	3 000	3 500	4 800	5 600
40	50	60	40	2 000	1 600	3 000	2 400
				£7 400	£7 200	£11 400	£11 150

Base weighted index of wages is $\dfrac{\Sigma p_n q_o}{\Sigma p_o q_o} \times 100$

$$= \frac{11\ 400}{7\ 400} \times 100 = 154.1$$

Current weighted index of wages is $\dfrac{\Sigma p_n q_n}{\Sigma p_o q_n} \times 100$

$$= \frac{11\ 150}{7\ 200} \times 100 = 154.9$$

Exercise 6.4

For the following data find the Laspeyres index and the Paasche index for prices for December 1971 with 1968 = 100.

	1968		*December 1971*	
	Price per unit (new pence)	*Value of purchases (new pence)*	*Price per unit (new pence)*	*Value of purchases (new pence)*
Bread	8.0	36.0	9.9	148.5
Butter	16.6	49.8	29.3	58.6
Jam	12.5	50.0	17.0	51.0

(I.O.S. Part I June 1973)

Calculation of index numbers using price relatives and weights

The Laspeyres and Paasche price index numbers assume that the quantities involved in the 'basket' of goods are easily identified. For food and drink there are no problems but in the case of housing there is no obvious unit of quantity. To get over the difficulties it is the practice to use as weights the *expenditure* on the item at base year.

As expenditure equals price × quantity, then if W is the weight then $W = p_o q_o$ and the sum of the weights $\Sigma W = \Sigma p_o q_o$.

When we want an index number of prices, we multiply the weights by the 'price relatives'. The price relatives are $\frac{p_n}{p_o}$. They are usually expressed as a percentage:

$$\frac{p_n}{p_o} \times 100\%$$

A price index number is then

$$\frac{\Sigma W \times \frac{p_n}{p_o}}{\Sigma W} \times 100$$

as $W = p_o q_o$, this equals $\frac{\Sigma p_o q_o \times \frac{p_n}{p_o}}{\Sigma p_o q_o} \times 100 = \frac{\Sigma p_n q_o}{\Sigma p_o q_o} \times 100$

If it is compared with earlier formulae, it will be noted that this is identical to a Laspeyres price index number.

Worked Example No. 6.5

General index of retail prices, January 1978

Group	*Weights*	*Price relatives (January 1974 = 100)*
Food	247	196.1
Alcoholic drink	83	188.9
Tobacco	46	222.8
Housing	112	164.3
Fuel and light	58	219.9
Durable household goods	63	175.2
Clothing and footwear	82	163.6
Transport and vehicles	139	198.7
Miscellaneous goods	71	198.6
Services	54	181.6
Meals bought and consumed outside the home	45	199.5

i Calculate the all items index for January 1978 with January 1974 = 100.

ii Calculate the index for non-food items for January 1978 with January 1974 = 100.

If we let the price relative $\frac{p_n}{p_o} \times 100 = I$

Then $\dfrac{\Sigma W \dfrac{p_n}{p_o} \times 100}{\Sigma W} = \dfrac{\Sigma WI}{\Sigma W}$

and this is similar in form to the formula for the mean. We can therefore use the same method as that given in Chapter 3.

Weight W	Index I	$W \times I$
247	196.1	48 436.7
83	188.9	15 678.7
46	222.8	10 248.8
112	164.3	18 401.6
58	219.9	12 754.2
63	175.2	11 037.6
82	163.6	13 415.2
139	198.7	27 619.3
71	198.6	14 100.6
54	181.6	9 806.4
45	199.5	8 977.5
1 000		190 476.6

Index number for all items $= \dfrac{\Sigma WI}{\Sigma W} = \dfrac{190\ 476.6}{1\ 000} = 190.5$

For an index number for non-food items we need to exclude the first and last items. We find $\Sigma W = 1\ 000 - 247 - 45 = 708$.

$\Sigma WI = 190\ 476.6 - 48\ 436.7 - 8\ 977.5 = 133\ 062.4$

Index number for non-food items $= \dfrac{\Sigma WI}{\Sigma W} = \dfrac{133\ 062.4}{708} = 187.9$

Exercise 6.5

Index of industrial production (1963 = 100)

Industry	*Weight*	*Index September 1970*
Mining and quarrying	56	76
Manufacturing industries:		
Food, drink and tobacco	84	129
Coal and petroleum products	7	149
Chemicals	64	160
Engineering	313	129
Metal manufacturing	60	114
Textiles	87	121
Bricks, pottery, etc.	29	130
Timber	21	130

Industry	*Weight*	*Index September 1970*
Paper and printing	59	122
Other manufacturing	25	159
Construction	127	115
Gas, electricity and water	68	116

Source: *Monthly Digest of Statistics*

Find the index of industrial production for September 1970:
a For all industries;
b For manufacturing industries.
Comment on your results.

Change of the base year

Most price index numbers in use today are base-year weighted. The difficulty with base-year weighting is that the quantity of goods we actually buy today may well be different from the quantity of goods we bought in the base year. The base year therefore has to be changed periodically. The index of retail prices has had base dates in 1947, 1952, 1956 and 1962. The present index is based on 16 January 1974 = 100.

If we want an index based on 16 January 1962 = 100, we have to 'splice' index numbers. This method is also called 'chaining' index numbers.

Worked Example No. 6.6

The index of retail prices for all items for January 1978 (January 1974 = 100) was found above to be 190.5. The index for January 1974 with January 1962 = 100 was 191.8.

January	1962	100	
January	1974	191.8	100
January	1978	Y	190.5

We need to find the index for January 1978 with January 1962 = 100.

From January 1974 to January 1978 the index increased from 100 to 190.5. We need to increase 191.8 in the same proportion, i.e. we need

$$\frac{Y}{191.8} = \frac{190.5}{100}$$

$$\text{Hence } Y = \frac{191.8 \times 190.5}{100} = 365.4$$

showing that the index of retail prices has increased by a factor of 3.654, or expressed in another way, by 265.4% since January 1962.

Exercise 6.6

Group	*Weight*	*Price relative August 1974 (Jan. 1974 = 100)*
Food	253	106.1
Alcoholic drink	70	110.7
Tobacco	43	120.3
Housing	124	105.1
Fuel and light	52	115.7
Durable household goods	64	109.5
Clothing and footwear	91	110.9
Transport and vehicles	135	112.7
Miscellaneous goods	63	113.3
Services	54	109.3
Meals bought and consumed outside the home	51	110.4

Source: *Department of Employment Gazette*

(i) Using the information given above, calculate the general index of retail prices for August 1974, with January 1974 = 100.

(ii) The previous base for this index was January 1962 = 100. With January 1962 as base, the figure for January 1974 was 191.8. Calculate the index for August 1974 with January 1962 as base.

Important note

The common form of examination questions on index numbers has now been covered. There are some other problems on index numbers that are occasionally set in examinations and these are covered briefly in the remainder of the chapter. Exercises on these topics are found at the end of the chapter.

Rebasing index numbers

Sometimes it is necessary to compare two index numbers, e.g. index of retail prices and index numbers of wholesale prices. The index of retail prices has January 1974 = 100 as a base and the index numbers of wholesale prices has average 1970 = 100 as a base. For comparative purposes it is necessary to rebase one of the index numbers so that both index numbers have the same base.

Worked Example No. 6.7

Monthly Averages	*Wholesale price index numbers, food manufacturing industries* (1970 = 100)	*General index of retail prices* (Jan 1962 = 100)	(Jan 1974 = 100)
1970	100	140.2	—
1971	109.2	153.4	—
1972	114.1	164.3	—
1973	132.6	179.4	—
1974	165.4	208.2	108.5
1975	199.8	—	134.8
1976	230.8	—	157.1

Source: *Annual Abstract of Statistics* 1977

We need to rearrange the general index of retail prices so that monthly averages 1970 = 100. The first step is to make the general index of retail prices into a single series (i.e. based on January 1962 = 100).

Let Y be the monthly average for 1975 with January 1962 = 100; then we have

$$\frac{Y}{208.2} = \frac{134.8}{108.5} \quad \therefore \quad Y = \frac{208.2 \times 134.8}{108.5} = 258.7$$

Similarly the monthly average for 1976 is

$$\frac{208.2 \times 157.1}{108.5} = 301.5$$

We now need to rebase the general index of retail prices with monthly average 1970 = 100; we do this as follows:

1971 $\quad \frac{153.4}{140.2} \times 100 = 109.4$

1972 $\quad \frac{164.3}{140.2} \times 100 = 117.2$

1973 $\quad \frac{179.4}{140.2} \times 100 = 128.0$

1974 $\quad \frac{208.2}{140.2} \times 100 = 148.5$

1975 $\quad \frac{258.7}{140.2} \times 100 = 184.5$

1976 $\quad \frac{301.5}{140.2} \times 100 = 215.0$

	Wholesale price index numbers (1970 = 100)	*General index of retail prices* (1970 = 100)
1970	100.0	100.0
1971	109.2	109.4
1972	114.1	117.2
1973	132.6	128.0
1974	165.4	148.5
1975	199.8	184.5
1976	230.8	215.0

Deflating index numbers

In the recent past earnings have increased rapidly but inflation has also been high. To find how real earnings have increased we need to 'deflate' the earnings statistics.

Worked Example No. 6.8

	General index of retail prices (*January 1974 = 100*)	*Monthly index of average earnings* (*January 1974 = 100*)
January	147.9	248.2
February	149.8	250.1
March	150.6	255.7
April	153.5	255.9
May	155.2	262.0
June	156.0	263.9
July	156.3	267.0
August	158.5	266.0
September	160.6	268.3
October	163.5	270.8
November	165.8	276.2
December	168.0	275.7

Deflate the monthly index of average earnings.

The method involves rebasing both series so that January 1976 = 100, i.e. we follow the method of Worked Example No. 6.7, then divide the rebased monthly index of average earnings by the corresponding rebased general index of retail prices and multiply by 100.

	General index of retail prices (Jan. 1976 = 100)	*Monthly index of average earnings* (Jan. 1976 = 100)	*Deflated monthly index of average earnings* Jan. 1976 = 100
	(i)	(ii)	(ii) ÷ (i) × 100
January	100.0	100.0	100.0
February	101.3	100.7	99.4
March	101.8	103.0	101.2
April	103.8	103.1	99.3
May	104.9	105.6	100.7
June	105.5	106.3	100.8
July	105.7	107.6	101.8
August	107.2	107.2	100.0
September	108.6	108.1	99.5
October	110.5	109.1	98.7
November	112.1	111.3	99.3
December	113.6	111.1	97.8

The deflated average earnings index shows that real earnings fell in 1976, particularly during the last four months.

Calculation of index numbers when data are given as expenditure

In national income accounting the data are usually given in two forms — expenditure at current prices and at base year prices. The method of finding index numbers in this situation is illustrated by the following worked example.

Worked Example No. 6.9

Consumers' expenditure (£ million)

	1970 at current prices	*1976 at current prices*	*1976 at 1970 prices*
Bread and cereals	822	1 672	779
Meat and bacon	1 786	3 895	1 740
Fish	193	421	178
Oil and fats	267	545	247
Sugar etc.	602	1 370	619
Fruit	365	756	362
Dairy products	930	1 959	996
Potatoes and vegetables	752	1 980	775
Beverages	403	1 000	512
Other	216	497	239
Total	6 336	14 095	6 448

Source: *National Income and Expenditure* 1977

Find a price index and a volume index for 1976 with 1970 = 100 for all the food items listed above.

As 1970 is base year, let $p_o q_o$ refer to 1970 prices and quantities and let $p_n q_n$ refer to 1976 prices and quantities.

1970 consumers expenditure at current (i.e. 1970) prices = $\Sigma p_o q_o$ = 6 336

1976 consumers expenditure at current (i.e. 1976) prices = $\Sigma p_n q_n$ = 14 095

1976 consumers expenditure at 1970 prices = $\Sigma p_o q_n$ = 6 448

A quantity index is $\dfrac{\Sigma p_o q_n}{\Sigma p_o q_o} \times 100 = \dfrac{6\,448}{6\,336} \times 100 = 101.8$

This is a *base* weighted index and is therefore a *Laspeyres index*.

A price index is $\dfrac{\Sigma p_n q_n}{\Sigma p_o q_n} \times 100 = \dfrac{14\,095}{6\,448} \times 100 = 218.6$

This is a *current* weighted index and is therefore a *Paasche index*.

NB: Expenditure has increased from 6 336 to 14 095.

$\dfrac{14\,095}{6\,336} \times 100 = 222.5$, i.e. expenditure for food has increased by 122.5%.

The two index numbers found above show that quantity of food consumed has increased by 1.8% and price of food has increased by

118.6%. Hence the increase in expenditure of 122.5% is made up of two components, an increase in quantity of 1.8% and an increase of price of 118.6%.

Please note: $\frac{101.8 \times 218.6}{100} = 222.5$

Revision Exercises

6.7 The wages paid to the manual workers on a company's pay roll are detailed below.

	1970 Average wage	*1970* Nos. employed	*1973* Average wage	*1973* Nos. employed
Men aged over 21	£30	350	£35	300
Women aged over 18	£22	300	£23	350
Youths	£15	80	£18	80
Girls	£10	60	£11	80

i With base year 1970 calculate for manual workers as a whole:

a a base weighted index number for wage rates;

b a current weighted index number for wage rates.

ii Account for the different results obtained in (a) and (b).

(B.A.B.S. 1973)

6.8 Calculate the index of industrial production for January 1976 using the data below.

i For all industries with 1970 = 100;

ii For all manufacturing industries with 1970 = 100;

iii For all industries with 1963 = 100, given that the index of industrial production for all industries for 1970 with 1963 = 100 was 124.2.

Index of industrial production, January 1976

Industry	*Weight*	*Index (1970 = 100)*
Mining and quarrying	37	87
Manufacturing:		
Food, drink and tobacco	84	103
Chemicals etc.	65	115
Metal	57	78
Engineering	319	99
Textiles	76	103
Other manufactures	144	103
Construction	146	91
Gas, electricity and water	72	112

Source: *Monthly Digest of Statistics*

(B.A.B.S. 1977)

6.9 External trade of the U.K.

Fresh vegetables	*1968* *Volume*	*Value*	*1969* *Volume*	*Value*
Potatoes	318	14.6	256	14.0
Onions	210	8.5	193	9.0
Tomatoes	167	24.2	158	25.2
Others	113	13.8	145	17.8

Volume in '000 tons
Value in £m

Source: *Annual Abstract of Statistics*

Compute for this group of imports an index number which gives the percentage change in price of imports from 1968 to 1969.

6.10 **a** Give the relative advantages and disadvantages of the Laspeyres and Paasche index numbers.

b For the following data find: (i) Laspeyres index; (ii) Paasche index for December 1971 with July 1968 = 100.

	July 1968		*December 1971*	
	Average price (new pence) per lb	*Average quantity purchased*	*Average price (new pence) per lb*	*Average quantity purchased*
Potatoes	1.8	3 lb	2.2	4 lb
Onions	4.5	1 lb	3.9	1 lb
Bacon	28.9	1 lb	38.7	$\frac{1}{2}$ lb
Ham	48.4	2 lb	58.7	3 lb
Cheese	21.6	1 lb	28.3	$1\frac{1}{2}$ lb
Sugar	3.5	5 lb	4.3	6 lb

(R.S.A. 1973)

6.11 **a** Briefly discuss the problems involved in comparing changes in retail prices over a long period.

b

Group	*Weights*	*Price relatives February 1976 (January 1974 = 100)*
Food	228	152.1
Alcoholic drink	81	150.9
Tobacco	46	162.8
Housing	112	135.8
Fuel and light	56	169.4
Durable household goods	75	141.2
Clothing and footwear	84	134.9
Transport and vehicles	140	156.9
Miscellaneous goods	74	154.2
Services	57	154.9
Meals bought and consumed outside house	47	148.3

Source: *Department of Employment Gazette*

i Calculate the general index of retail prices for all items for February 1976 with January 1974 = 100.

ii A pensioner purchased a £500 index-linked savings certificate in July 1975 when the general index of retail prices was 138.5. The value of index-linked savings certificates increases with the percentage change in the general index of retail prices. Calculate the value of the savings certificate in February 1976.

(B.A.S.S. (P/T) 1976)

6.12 Consumers' expenditure (£m)

	Food		*Housing*	
	At current prices	*At 1970 prices*	*At current prices*	*At 1970 prices*
1970	6 365	6 365	4 181	4 181
1971	6 964	6 362	4 719	4 287
1972	7 423	6 320	5 432	4 377
1973	8 443	6 388	6 305	4 438
1974	9 837	6 418	7 198	4 460

Calculate *price* index numbers for all the years given for **a** food and **b** housing.

State what type of index number you have constructed and what you understand by this.

Use your results to compare price movements in the two sectors over the given time period, using a graph if you wish.

(H.N.D. 1977)

6.13 Consumers' expenditure in U.K. (£m)

	At current prices	*Revalued at 1970 prices*
1970	31 387	31 387
1971	34 881	32 241
1972	39 472	34 179
1973	44 855	35 759

i Explain the purpose of publishing these figures both at current prices and at a fixed base price.

ii From the data obtain price *and* volume index numbers, indicating what type of index numbers you have obtained.

(H.N.D. 1975)

6.14 Index numbers

	(1) *Average earnings* (Jan 1970 = 100)	(2) *Retail price index* (Jan 1974 = 100)
1976		
May	259.6	155.2

	(1) *Average earnings* (Jan 1970 = 100)	(2) *Retail price index* (Jan 1974 = 100)
June	261.2	156.0
July	263.1	156.3
August	267.2	158.5
September	266.1	160.6
October	269.0	163.5
November	272.2	165.8
December	277.1	168.0
1977		
January	278.1	172.4
February	278.7	174.1
March	283.8	175.8
April	283.1	180.3
May	286.3	181.7

a Convert these two series of index numbers to a base of May 1976 = 100 and plot on the same graph.

b Comment on your results.

(H.N.D. 1978)

6.15 Using the results of question 6.14, give the average earnings index deflated by the retail price index to a base of May 1976 = 100 and comment on your results.

7 Time Series

A time series is obtained when we are given successive values of a variable at successive intervals of time. Many economic indicators, index of retail prices, overseas trade statistics, index of industrial production, retail sales, etc. — are published as a time series. Most forecasting techniques involve the use of time series data.

In a simple analysis, a time series is assumed to contain the following components.

i *The trend*: the longer-term underlying movement of the series.

ii *The cyclical component*: wave-like movement caused for example by the effects of a 'stop-go' economy.

iii *The seasonal component*: a yearly cycle caused by the seasonal influence of spring, summer, autumn, winter, which affects many business and economic time series.

iv *The residual or 'disturbance' component*: This is taken to be a random, non-systematic component, affecting the time series through such unpredictable events as strikes, breakdowns of plant, non-seasonal illness, bad weather etc.

Included in category (iv) are factors which have an extremely severe effect on the economy and the corresponding time series, viz. major strikes — the miners' strikes in 1972 and 1974; weather — the bad winter of 1963; wars — the Arab-Israeli conflicts. Special techniques have been devised to decide whether such factors are extreme and to handle the factors that are deemed extreme. A simple method of dealing with extreme items may be illustrated by considering the output of mining and quarrying in the first quarter of 1972 (during the miners' strike). The actual output could be replaced by the average of the output in the first quarter of 1971 and the output in the first quarter of 1973.

In an elementary analysis we usually combine categories (i) and (ii), the trend and cyclical component, and these are simply called the trend. We also assume that extreme items have already been dealt with. Thus in an elementary analysis we assume that the series has three components; trend, seasonal component, and residual component.

Figures 7.1 to 7.4 illustrate typical time series. Figure 7.1 shows a series with an upward trend. Figure 7.2 shows a series with a pronounced seasonal variation but with random influences; the high peaks of 1975 and 1976 reflect the good summer weather of these two years and can be regarded as random components of the time series. Figure 7.3 shows an upward trend with cyclical factors. Figure 7.4 shows a downward trend but with large random factors, the effect of the miner's strikes of 1972 and 1974; these are regarded as extreme random elements.

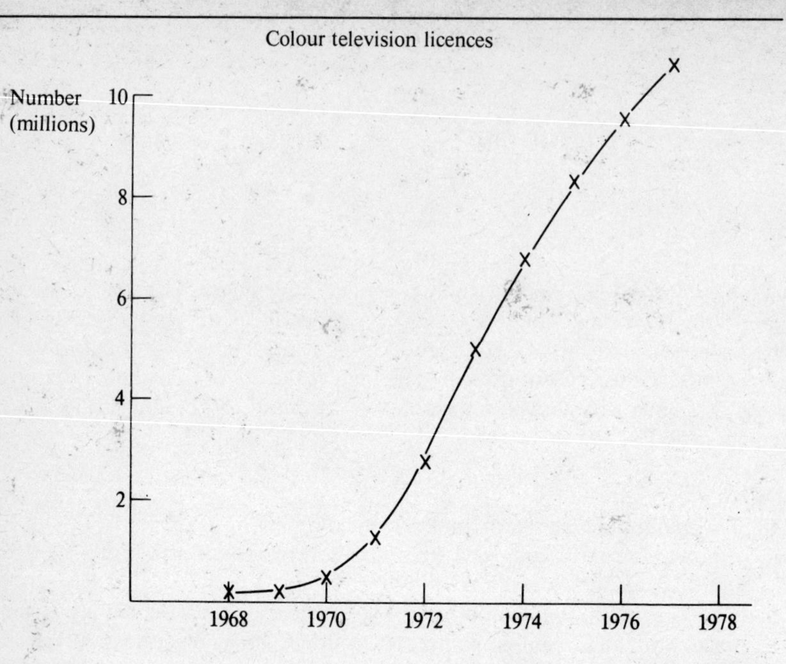

Figure 7.1

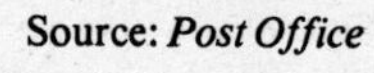
Source: *Post Office*

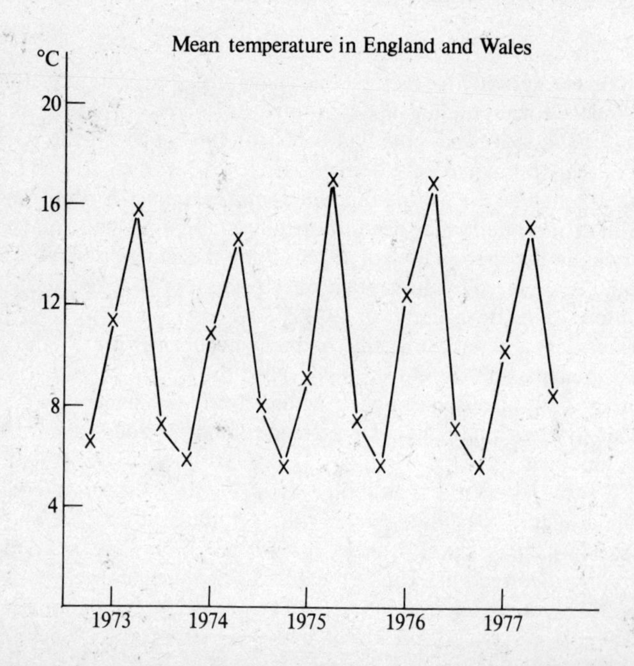

Figure 7.2 Source: *Meteorological Office*

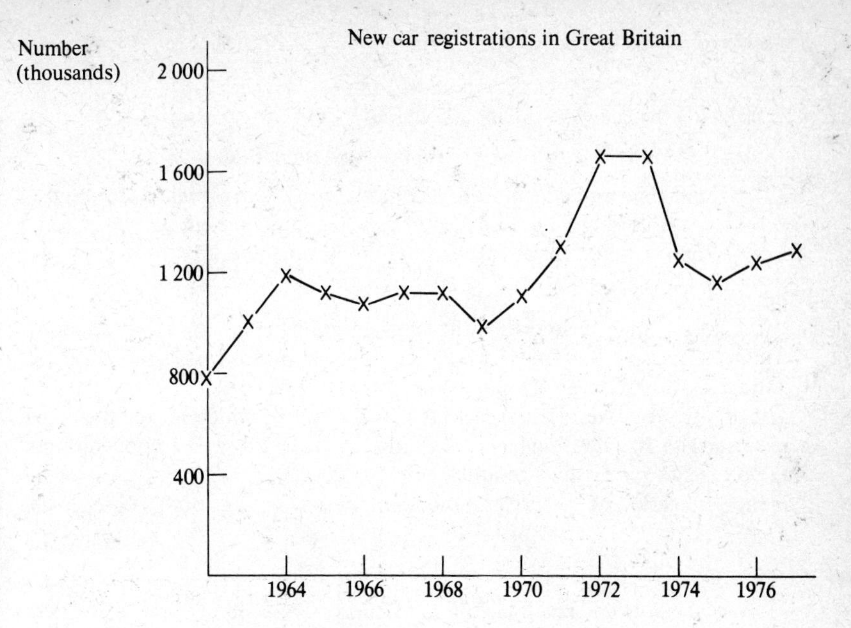

Figure 7.3

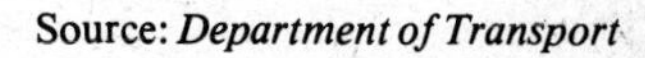
Source: *Department of Transport*

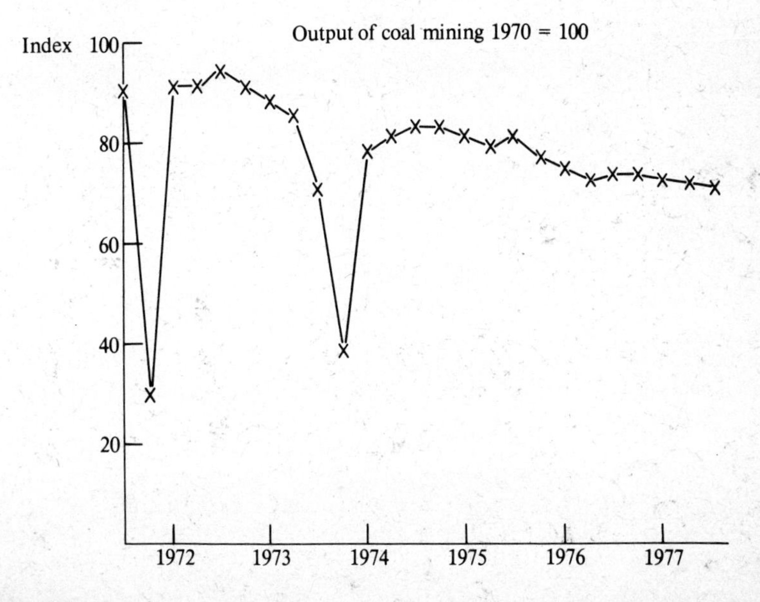

Figure 7.4 Source: *Central Statistical Office*

The additive model

This assumes that any observed value of the time series is the result of adding (algebraically) the three components, yielding the following relationship:

Observed value = Trend *plus* seasonal *plus* residual factors.

If Y represents an observation of the time series, T represents the trend, S represents the seasonal factor, and R represents the random element, then we may express the above statement as follows:

$$Y = T + S + R$$

The seasonal component is thought of as oscillating in a cyclical pattern about the trend. A negative sign for the seasonal movement means that it is below trend. A positive seasonal movement means that it is above trend.

Similarly, a negative residual reflects some influence which makes the actual observation less than it would have been with trend and seasonal components only, and conversely for a positive residual.

In the application of the additive model the following assumptions are usually made:

- i Trend, seasonal and residual components act independently of one another.
- ii Seasonal pattern remains stable year after year, e.g. if S_1, S_2, S_3, S_4 are the seasonal components for the four quarters of the year, then these components have the same value for each of the years and the following years.
- iii $\Sigma S_i = 0$ for quarterly data (i.e. the 4 values of S add up to zero, or cancel each other out over a year.)

The additive model is illustrated in Figure 7.5.

The multiplicative model

It is equally permissible to assume a multiplicative model, that is to regard an actual observation as the *product* not the sum of the three components, yielding the formulation:

Observed value = Trend × Seasonal × Residual

This may be expressed as follows:

$$Y = T \times S \times R$$

The multiplicative model should always be preferred to the additive model if the seasonal fluctuation appears to change proportionally to the trend in the data. In the multiplicative model, seasonality is expressed in percentage rather than absolute terms; for example, a seasonal index of 110% implies that the seasonal component is carrying the series above the trend and in absolute terms is 10% of the corresponding trend value. We assume that:

- i The seasonal indices remain constant year after year.

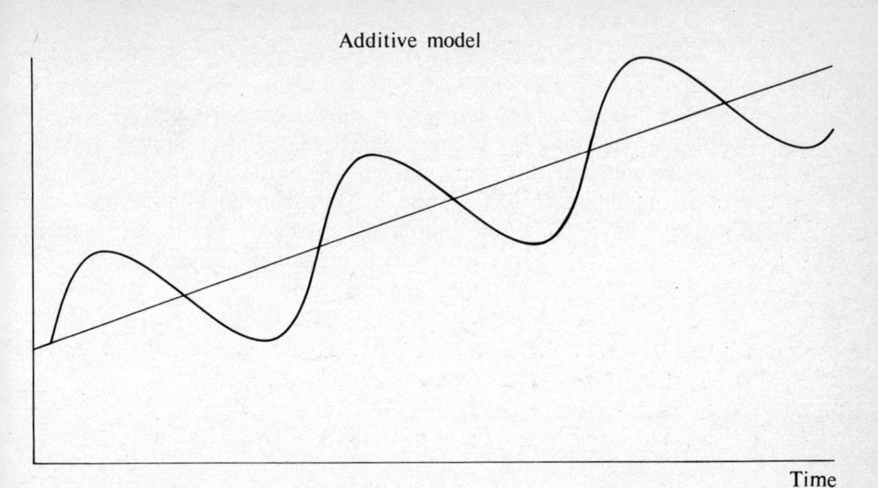

Figure 7.5

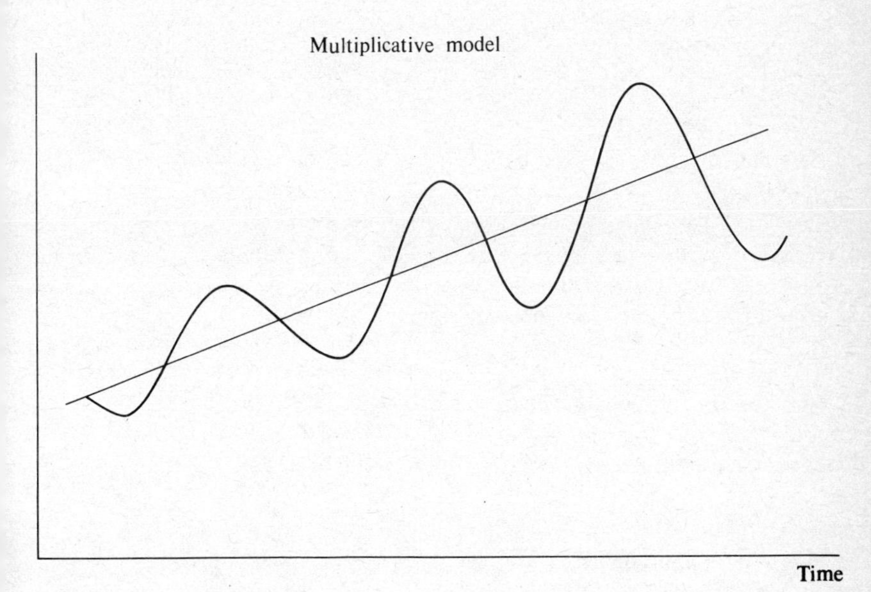

Figure 7.6

ii $\Sigma S_i = 400$ for quarterly data (i.e. the 4 values of S have a mean of 100%)

The multiplicative model is illustrated in Figure 7.6.

The trend

The standard method of finding the trend is to use a moving average. Suppose we have yearly data of new car registrations (the data illustrated in Figure 7.3). We can use a 3-year, 5-year, or 7-year moving average. The period of the moving average should be approximately equal to the periodicity of the data. (In some examinations questions you are asked to find, say, a 5 term moving average. In other questions you can be asked to use an appropriate moving average; in this situation you need to plot the data to find the periodicity of the data and hence the number of terms to use in the calculation of the moving average.)

Worked Example No. 7.1

New car registrations in Great Britain (thousands)

Year	*No. of registrations*	*Year*	*No. of registrations*
1962	786	1970	1 099
1963	1 011	1971	1 304
1964	1 193	1972	1 676
1965	1 125	1973	1 658
1966	1 068	1974	1 247
1967	1 119	1975	1 180
1968	1 120	1976	1 271
1969	990	1977	1 288

Source: *Department of Transport*

Find the trend for the above data.

From Figure 7.3 there is no clear periodicity but a 5-year moving average is used in this example. The first step is to add the first five terms, i.e. for 1962–1966, 786 + 1 011 + 1 193 + 1 125 + 1 068 = 5 183. This is placed in column 3. The next step is to divide by 5 i.e. 5 183 ÷ 5 = 1 036.6. This is placed in column 4. To find the next term in the moving average we take the years, 1963–1967 i.e. 1011 + 1 193 + 1 125 + 1 068 + 1 119 = 5 516 and divide by 5 i.e. 5 516 ÷ 5 = 1 103.2. Similarly for the remaining terms.

Important note

As a check on arithmetical accuracy it is useful to note that 5 516 = 5 183 + 1 119 − 786, 5 625 = 5 516 + 1 120 − 1 011 and so on until you reach 6 644. If you now check that 1973–1977 total to 6 644, i.e. 1 658 + 1 247 + 1 180 + 1 271 + 1 288 = 6 644, it is reasonable to assume that all moving totals are now correct.

It should be noted that in the moving average there are two gaps, at the beginning and the end.

Table 1

(thousands)

Year	*No. of car registrations*	*5-year total*	*5-year moving average*
1962	786	—	—
1963	1 011	—	—
1964	1 193	5 183	1 036.6
1965	1 125	5 516	1 103.2
1966	1 068	5 625	1 125.0
1967	1 119	5 422	1 084.4
1968	1 120	5 396	1 079.2
1969	990	5 632	1 126.4
1970	1 099	6 189	1 237.8
1971	1 304	6 727	1 345.4
1972	1 676	6 984	1 396.8
1973	1 658	7 065	1 413.0
1974	1 247	7 032	1 406.4
1975	1 180	6 644	1 328.8
1976	1 271	—	—
1977	1 288	—	—

Exercise 7.1

The profits of a company over the 14-year period 1965–1978 are given below.

Year	*Profits* (£1 000)	*Year*	*Profits* (£1 000)
1965	149	1972	312
1966	182	1973	301
1967	219	1974	337
1968	200	1975	380
1969	205	1976	339
1970	231	1977	356
1971	281	1978	397

Find the 5-year moving average for the above data.

Method of finding the trend when the data is given for quarters

In the case of quarterly data it is essential that the moving average should be centred appropriately. Suppose we have the following data:

Worked Example No. 7.2

Sales revenue

Year		*1975*	*1976*	*1977*	*1978*
Quarters:	1 Spring	79	97	113	136
	2 Summer	48	66	91	105
	3 Autumn	68	85	100	125
	4 Winter	107	134	148	174

Here we have data with an obvious seasonality. Thus the periodicity is 4 and the moving average should be based on 4 terms. However the moving average for the first 4 terms would be centred on the middle of the year i.e: relate to 30 June 1975. To deal with seasonal factors, we need the moving average to be centred at the middle of the quarter. To achieve this we take the first two terms of the 4 quarter moving average, these would be centred on 30 June 1975 and 30 September 1975, and then average these two terms. This average would be centred midway between 30 June 1975 and 30 September 1975, i.e: the middle of the third quarter. The most simple way of achieving this is as follows (see Table 2 on page 120):

Calculate the four term moving totals, then add these moving totals in pairs and then divide this total by 8. The method is illustrated in columns (i) to (iv) of example 7.2. Column (iv) gives a centred moving average. This centred moving average is in fact a 5-term moving average with weights 1, 2, 2, 2, 1. The moving average is an estimate of the trend.

Seasonal movements

Before attempting to find the seasonal movements it is necessary to discover whether the model is additive or multiplicative. If the data are additive, then the amplitude (the extent of oscillation) of the series about the trend curve should be approximately constant, as in Figure 7.5. If the data is multiplicative, then the amplitude of the series about the trend curve is approximately proportional to the trend, as in Figure 7.6. A plot of the data would be the best method but in the examination room this would take time. A quick method is to take differences between the smallest value and the largest value for each year. If these differences are approximately equal, then the additive model is appropriate; if the differences are increasing (assuming an upward trend), then the multiplicative model is appropriate.

For worked example No. 7.2 we find

Year	*Largest*	—	*Smallest*		*Difference*
1975	107	—	48	=	59
1976	134	—	66	=	68
1977	148	—	91	=	57
1978	174	—	105	=	69

As the differences are approximately equal an additive model is appropriate.

In example 7.2 therefore we shall assume the additive model, i.e.

Observed value = Trend + Seasonal factor + Residual factor.

If we subtract the trend from the observed value we obtain:

Seasonal factor + Residual factor.

In our example col. (iv) is subtracted from col. (i) to give col. (v), the seasonal factor + residual factor for each quarter. To obtain the seasonal factor we take the average of the elements in col. (v) for each quarter. This process is carried out in Table 2(a). When we average the seasonal factor + residual factor for each quarter we assume that the average of the residual factors is zero and hence this resulting average should give the seasonal factor for each quarter.

It is necessary to ensure that the sum of the seasonal factors should be zero. Table 2(a) therefore includes an adjustment to enable this to be attained.

In col. (vi) we place the seasonal factors as found from Table 2(a). We then subtract algebraically the seasonal factors from the original data. We subtract col. (vi) from col. (i) — the results are given in col. (vii), which provides the seasonally adjusted values of the data, i.e. the value of the data with the seasonal effect eliminated. It is important to recognise that the seasonally adjusted values do not give the trend. The seasonally adjusted values are in fact equal to the trend plus the residual component. The trend is given by col. (iv). An estimate of the residual components can be obtained if we subtract the trend from the seasonally adjusted series, i.e. subtract col. (iv) from col. (vii).

Important note

In examination questions you are often asked to find the trend. Column (iv) should be clearly labelled 'Trend' to show the examiner that you have found the trend.

Table 2

Year and quarter		*Data*	*Add in fours*	*Add col (ii) in twos*	*Trend: divide Col (iii) by 8*	*Col.* (i) – *Col.* (iv)	*Seasonal variation (see Table 2(a) below)*	*Seasonally adjusted data* (Col. (i)–Col. (vi))
		(i)	(ii)	(iii)	(iv)	(v)	(vi)	(vii)
1975	1	79					+9	70
	2	48	302				−24	72
	3	68	320	622	77.8	−9.8	−13	81
	4	107	338	658	82.3	+24.7	+28	79
1976	1	97	355	693	86.6	+10.4	+9	88
	2	66	382	737	92.1	−26.1	−24	90
	3	85	398	780	97.5	−12.5	−13	98
	4	134	423	821	102.6	+31.4	+28	106
1977	1	113	438	861	107.6	+5.4	+9	104
	2	91	452	890	111.3	−20.3	−24	115
	3	100	475	927	115.9	−15.9	−13	113
	4	148	489	964	120.5	+27.5	+28	120
1978	1	136	514	1003	125.4	+10.6	+9	127
	2	105	540	1054	131.8	−26.8	−24	129
	3	125					−13	138
	4	174					+28	146

Table 2a

Quarter	1	2	3	4
Year 1975	—	—	− 9.8	+24.7
1976	+10.4	−26.1	−12.5	+31.4
1977	+ 5.4	−20.3	−15.9	+27.5
1978	+10.6	−26.8	—	—
Total	+26.4	−73.2	−38.2	+83.6
Average	+ 8.8	−24.4	−12.7	+27.9
Adjustment	+ 0.1	+ 0.1	+ 0.1	+ 0.1
Seasonal factor	+ 8.9	−24.3	−12.6	+28.0
Rounded seasonal factor*	+ 9	−24	−13	+28

* Seasonal factors are given to the same accuracy as the original data.

Seasonal factors must be adjusted to sum to zero: $+8.8 - 24.4 - 12.7 + 27.9 = -0.4$

Add $\frac{+0.4}{4} = +0.1$ to each average.

Exercise 7.2

U.K. passenger movements by sea (100 000)

Quarter / *Year*	*1*	*2*	*3*	*4*
1973	6	17	33	8
1974	6	19	34	10
1975	9	21	39	10

Source: *Monthly Digest of Statistics*

a By means of a moving average, find the trend and the seasonal adjustments.

b Give the passenger movements for 1975 seasonally adjusted.

(B.A.S.S. (P/T) 1976)

Important note

This chapter includes both additive and multiplicative models. Many examiners will accept either model and no loss of marks would occur if you happened to choose the wrong model. However, if the question specifically asks you to select the appropriate model, you would lose some marks if you selected the wrong model.

If you are not allowed a calculator in the examination room the additive model is recommended simply because the arithmetic is less. Many examination candidates in fact learn only the additive model. A survey of past examination papers would show you whether it is safe to omit the multiplicative model.

Multiplicative model

In Worked Example 7.3 we shall consider a multiplicative model, i.e.

Observed values = Trend × Seasonal factor × Residual factor.

If we *divide* the observed value by the trend we obtain

Seasonal factor × Residual factor

Cols. (i) to (iv) of Table 3 are the same as for the additive model, but col. (v) is obtained by dividing col. (i) by col. (iv) and then multiplying by 100 to obtain a percentage value. To obtain the seasonal factor we take the average of the elements of col. (v) for each quarter. This process is carried out in Table 3a. It is necessary to ensure that the sum of the seasonal factors (or seasonal indices as they are sometimes called) is 400; Table 3a therefore includes an adjustment to enable this to be attained.

Worked Example No. 7.3

Sales of fertilizer (thousands of tons)

Quarter	*1973*	*1974*	*1975*	*1976*	*1977*
1st	48	50	68	93	84
2nd	52	46	34	56	61
3rd	16	22	26	16	29
4th	35	40	35	45	48

By means of a moving average find the trend and, assuming a multiplicative model, find the seasonal indices for each quarter.

It should be noted that the difference between the largest and smallest observation for each year increases approximately with the trend.

Year	*Largest*	—	*Smallest*		*Difference*
1973	52	—	16	=	36
1974	50	—	22	=	28
1975	68	—	26	=	42
1976	93	—	16	=	77
1978	84	—	29	=	55

Table 3

Year and quarter		*Data* (i)	*Add in fours* (ii)	*Add col. (ii) in twos* (iii)	*Trend: divide col. (iii) by 8* (iv)	*Col. (i) ÷ Col. (iv) as a percentage* (v)	*Seasonal indices (from Table 3a)* (vi)	*Seasonally adjusted data — Col. (i) × 100 ÷ Col. (vi)* (vii)
1973	1	48					159.9	30
	2	52	151				105.8	49
	3	16	153	304	38.0	42.1	47.0	34
	4	35	147	300	37.5	93.3	87.3	40
1974	1	50	153	300	37.5	133.3	159.9	31
	2	46	158	311	38.9	118.3	105.8	43
	3	22	176	334	41.8	52.7	47.0	47
	4	40	164	340	42.5	94.1	87.3	46
1975	1	68	168	332	41.5	163.9	154.9	43
	2	34	163	331	41.4	82.2	105.8	32
	3	26	188	351	43.9	59.3	47.0	55
	4	35	210	398	49.8	70.4	87.3	40
1976	1	93	200	410	51.3	181.5	159.9	58
	2	56	210	410	51.3	109.3	105.8	53
	3	16	201	411	51.4	31.1	47.0	34
	4	45	206	407	50.9	88.5	87.3	52

Year and quarter		*Data*	*Add in fours*	*Add col. (ii) in twos*	*Trend: divide col. (iii) by 8*	*Col. (i) ÷ Col. (iv) as a percentage*	*Seasonal indices (from Table 3a)*	*Seasonally adjusted data — Col. (i) ×100 ÷ Col. (vi)*
		(i)	(ii)	(iii)	(iv)	(v)	(vi)	(vii)
1977	1	84	219	425	53.2	158.1	159.9	53
	2	61	222	441	55.1	110.7	105.8	58
	3	29					47.0	62
	4	48					87.3	55

Table 3a

Quarter	1	2	3	4
Year				
1973	—	—	42.1	93.3
1974	133.3	118.3	52.7	94.1
1975	163.9	82.2	59.3	70.4
1976	181.5	109.3	31.1	88.5
1977	158.1	110.7	—	—
Total	636.8	420.5	185.2	346.3
Average	159.2	105.1	46.3	86.6
Adjustment	0.7	0.7	0.7	0.7
Seasonal factor	159.9	105.8	47.0	87.3

Note: Seasonal indices must be adjusted to sum to 400.
$159.2 + 105.1 + 46.3 + 86.6 = 397.2$

i.e. $400 - 397.2 = 2.8 \quad \frac{2.8}{4} = 0.7$

Add 0.7 to each.

Exercise 7.3

Sales of gas (100 million therms)

Year/quarter	*1*	*2*	*3*	*4*
1972	29	22	17	29
1973	33	23	18	33
1974	41	27	22	37
1975	44	28	21	

Source: *Monthly Digest of Statistics*

By means of a moving average, find the trend and hence estimate the seasonal components.

(I.O.S. Nov. 1976)

Forecasting techniques

Most forecasting techniques involve projecting a trend forward. Suppose that in Example 7.2 we were asked to forecast sales revenue for each quarter of 1979. The difficulty in making a forecast is in whether to project the future trend forward — this assumes that the trend will continue as in the past.

Worked Example No. 7.4

We plot the trend values from Col. (iv) of Table 2 on graph paper. In the examination room you have only time to fit the line of best fit by eye (more sophisticated methods are available but are not expected in most examinations at which this book is aimed). This is shown in Figure 7.7. The dotted line represents the trend and this projected into the future allows us to estimate future trend values. The *trend* estimates for 1979 are: 1st Quarter 145, 2nd Quarter 150, 3rd Quarter 155, 4th Quarter 160. Forecast of sales revenue for each quarter is the estimated trend value *plus* seasonal adjustment.

		Trend	*Seasonal adjustment*	*Forecast*
1979	1	145	+9	154
	2	150	-24	126
	3	155	-13	142
	4	160	+28	188

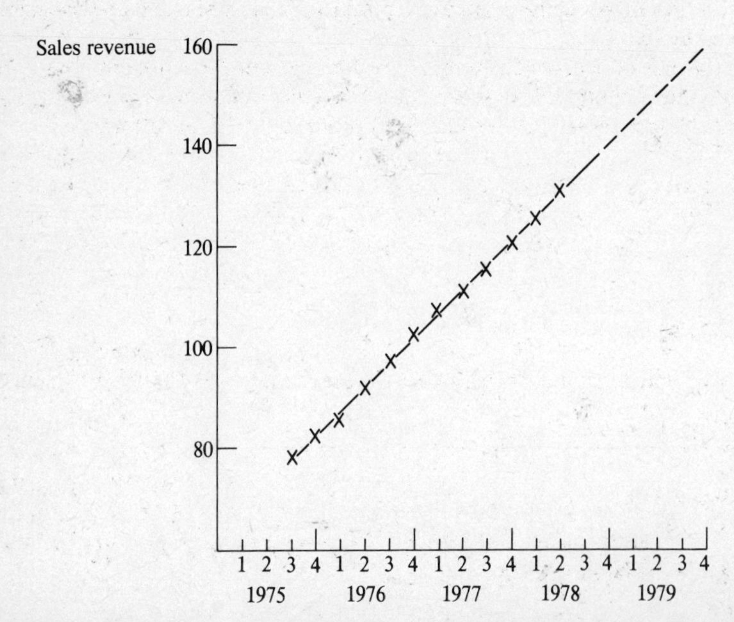

Figure 7.7

Forecasts are frequently inaccurate. The method illustrated above assumes:

- **i** That a straight line is a good fit to past trend values;
- **ii** That the trend line when projected forward represents the future;
- **iii** That the seasonal adjustments of the past are applicable in the future;
- **iv** That the model (in this case additive) will be applicable in the future.

Furthermore we cannot anticipate random influences, e.g. if the product we are selling is electric blankets, sales would be higher if we had a very severe winter.

Exercise 7.4

Using the data of Exercise 7.2, forecast U.K. passenger movements for each quarter of 1976.

Forecasting using the multiplicative model

The forecast in this case is equal to the product of the trend and the seasonal index divided by 100.

In the case of Worked Example No. 7.2 it is possible to work out the seasonal index using a multiplicative model. If we take the trend values already found in Worked Example No. 7.4, we find the forecast values in the following table:

Trend value col. (i)	*Seasonal index* col. (ii)	*Forecast value* col. (i) × col. (ii) ÷ 100
145	108.3	157
150	77.5	116
155	86.8	135
160	127.4	204

It will be noted that a different model produces a markedly different forecast.

Exercise 7.5

Using the data of Exercise 7.3 forecast the sales of gas in the fourth quarter of 1975.

Revision Exercises

7.6 **a** Briefly describe the components which make up a time series.

b **Sales of caravans (thousands)**

Quarters	*1*	*2*	*3*	*4*
1973	40	37	25	31
1974	26	30	19	21
1975	25	25	14	18

Source: *Monthly Digest of Statistics*

i By means of a moving average, find the trend and the seasonal adjustments.

ii Give the sales of caravans for 1975 seasonally adjusted.

(H.N.D. 1976)

7.7 **Houses completed in U.K. (thousands)**

	For local authority	*Private*
1960	128	171
1961	116	180
1962	129	178
1963	124	178
1964	155	221
1965	165	217
1966	177	209
1967	200	204
1968	188	226
1969	181	186
1970	177	174
1971	155	196
1972	120	201
1973	103	191

By means of a five-year moving average, find the trend for both sets of data. Comment on your results.

(R.S.A. 1975)

7.8 **i** Explain briefly why time series data may need 'adjusting' for the seasonal component in that data.

ii Make an estimate of the seasonal component in the following figures.

Index nos. of retail sales in department stores
(Weekly average 1971 = 100)

	1st quarter	*2nd quarter*	*3rd quarter*	*4th quarter*
1971	83	91	95	131
1972	93	104	112	154
1973	114	116	124	168

(H.N.D. 1975)

7.9 **a** Indicate the situations in which you would employ (i) an additive model; and (ii) a multiplicative model, to estimate seasonal variations in time series analysis.

b What is the value of sales forecasting to a company?

c **Sales of carpets**
(10 000 square metres)

Quarter	*1*	*2*	*3*	*4*
1973	—	—	—	455
1974	393	416	368	440
1975	402	415	382	470
1976	454	435	415	—

Source: *Monthly Digest of Statistics*

By means of a moving average, find the trend and the seasonal adjustments and hence forecast sales for the fourth quarter of 1976.

7.10 **Air departures from the United Kingdom for Europe and the Mediterranean, 1960–1964**

		Departures (thousands)			*Departures (thousands)*
1960	1st quarter	258	1962	1st quarter	327
	2nd quarter	595		2nd quarter	792
	3rd quarter	879		3rd quarter	1 191
	4th quarter	339		4th quarter	406
1961	1st quarter	330	1963	1st quarter	354
	2nd quarter	700		2nd quarter	890
	3rd quarter	1 075		3rd quarter	1 324
	4th quarter	372		4th quarter	462

By means of a moving average find the trend and the seasonal component in the series. Hence obtain estimates of departures for each of the first two quarters of 1964.

The actual departures for the first and second quarters of 1964 were 440 000 and 946 000 respectively. Compare your estimates with the actual figures and comment.

(B.Sc.(Econ). 1965)

7.11 **Marriages in England and Wales, 1975–1977**
(*thousands*)

Quarter	*1*	*2*	*3*	*4*
1975	78	95	127	85
1976	82	88	127	83
1977	74	85	117	82

Source: *Monthly Digest of Statistics*

i By means of a moving average find the trend and the seasonal adjustments.
ii Give the data for 1977 seasonally adjusted.
iii If you were asked to forecast the number of marriages for each quarter of 1978, outline how you would attempt to do this and discuss the likely accuracy of your forecast.

(B.A.S.S. (P/T) 1978)

7.12 Live births in England and Wales 1974–1976 (thousands)

Quarter	*1*	*2*	*3*	*4*
1974	162	163	164	150
1975	155	156	153	140
1976	151	150	147	137

Source: *Monthly Digest of Statistics*

a By means of a moving average find the trend and the seasonal adjustments.
b Forecast the number of live births in England and Wales for the first two quarters of 1977.
c Briefly discuss the problems in forecasting the number of births.

(B.A.S.S. (P/T) 1977)

8 Probability

Important note

It is essential for an understanding of Chapter 9, 10 and 11 to appreciate the basic ideas and principles of probability. Questions set on the contents of these chapters have, however, a relatively small probability content. In examinations questions are often set which are entirely related to probability. As a teacher, as an examiner and as an assessor of statistics examinations, I am aware that such probability questions are not very popular with candidates. Those candidates who do attempt probability questions fall into two categories: a small number of candidates who are very good at probability and score almost full marks; and the remainder, who do very badly and score few, if any, marks.

This chapter covers most of the standard questions set on probability at the elementary level. If you have difficulty in doing the exercises, do not worry; many students have difficulty with probability but still pass a statistics examination with ease. A word of warning to those who find the exercises of this chapter easy: an examiner, by a slight change of wording, can set a question which is not specifically covered in the worked examples and exercises; if a candidate does not appreciate the effect of the slight change of wording he may do the question very badly.

While it is essential to read the chapter to gain an understanding of the ideas and principles of probability, I **do not recommend that candidates attempt probability questions in the examination room**. However, if you have done a question **exactly** like the one set and are confident that you can solve the problem, you could perhaps attempt a probability question. In general it is much better to try other questions and hope to score say 10 marks out of 20 than risk a probability question when you might score no marks. In my experience as an examiner many candidates have lost marks, and some, as a result failed simply because they attempted a probability question and scored virtually nothing on the question, whereas another question would have yielded them a few marks.

Subjective probability

The first contact we have with probability is usually in the field of gambling. If a bookmaker offers odds of 9 to 1 against a horse winning a race and he accepts a bet of £1, then he keeps the £1 if the horse loses and pays out £9 and returns the bet of £1 if the horse wins. The bookmaker believes that there is a probability of at least $\frac{9}{9+1} = \frac{9}{10}$, i.e. 9 chances out of 10, that the horse will lose and a probability of not more than $\frac{1}{9+1} = \frac{1}{10}$, i.e. 1 chance in 10, that the horse will win. This type of probability is based on *the degree of belief* that an event will occur. This type of probability is usually called *subjective* probability. An example of subjective probability in the marketing field is given below.

A firm is considering launching a new product. The state of the economy at the time of the launch is thought to affect the likely success of the product. Suppose

A is the event that national income next year will rise by more than 5%.

B is the event that national income next year will rise by 5% or less.

C is the event that national income next year be the same as this year.

D is the event that national income next year will fall.

The marketing manager will consider the various forecasts of the economy for next year, taking into account past knowledge of the reliability (or perhaps unreliability!) of such forecasts. He may think that the probability that event A — national income next year will rise by more than 5% — will occur is $\frac{1}{4}$. We may write this sentence in symbols as follows:

$$P(A) = \tfrac{1}{4}$$

P(A) means the probability that event A will occur. Similarly P(B) means probability that event B will occur.

The marketing manager may also think that $P(B) = \frac{3}{8}$, $P(C) = \frac{1}{8}$, $P(D) = \frac{1}{4}$. In numerical terms probability must lie between 0 (no chance of event occurring) and 1 (certainty of event occurring). In this example, clearly one of the events A, B, C, D must occur. Thus

$$P(A) + P(B) + P(C) + P(D) = 1. \quad \tfrac{1}{4} + \tfrac{3}{8} + \tfrac{1}{8} + \tfrac{1}{4} = 1$$

Classical probability

Probability problems involving rolling dice, drawing cards from a pack of playing cards, drawing coloured balls from a bag are frequent topics for examination questions. This is usually called 'classical' probability. In this situation the probability of the outcome can usually be determined before the event takes place. Consider the following examples.

1. If we roll a true six sided dice the possible outcomes are 1, 2, 3, 4, 5 or 6 and each outcome is equally likely. The probability of rolling a six $P(6) = \frac{1}{6}$.
2. If a card is drawn from a pack of 52 playing cards, what is the probability that the card is (i) the ace of clubs (ii) not the ace of clubs? Let A be the event 'ace of clubs', then by convention the event 'not ace of clubs' is denoted by $\overline{A}$. As there is only one ace of clubs in a pack of playing cards, $P(A) = \frac{1}{52}$, $P(\overline{A}) = \frac{51}{52}$. It will be noted that $P(A) + P(\overline{A}) = 1$.

Exercise 8.1

A card is drawn from a pack of playing cards. What is the probability that the card is (i) an ace, (ii) a club, (iii) a court card, i.e. king, queen or jack?

Relative frequency basis of probability

Suppose that a six-sided dice is biased. In this case it is unlikely that the probability of rolling a six P(6) = $\frac{1}{6}$, the result we would expect from classical probability. To find P(6) we could roll the dice 3 000 times and count the number of sixes. If we observe 600 sixes, then

$$P(6) = \frac{\text{Number of sixes observed}}{\text{Number of trials}} = \frac{600}{3\,000} = \frac{1}{5}$$

This form of probability is based on the relative frequency of the event occurring and is the form of probability usually employed by statisticians.

Mutually exclusive outcomes

When there are several possible outcomes of an action and it is impossible for more than one outcome to occur at any one time, these outcomes are said to be mutually exclusive.

In the case of the example on marketing discussed on page 130, four possible outcomes of the national income were listed: A (a rise of more than 5%), B (a rise of 5% or less), C (no change), D (a fall). Clearly one and only one of these outcomes will occur; thus events A, B, C and D are mutually exclusive.

Additive law of probability

If events are *mutually exclusive*, then the probability that two or more outcomes will occur is the sum of the separate probabilities. In symbols this means

$$P(A + B) = P(A) + P(B)$$

In the case of mutually exclusive events, P(A + B) means the probability that A or B will occur.

In our example on marketing discussed on page 130 the probability that national income will rise next year P(R) is the probability that events A or B will occur thus

$$P(R) = P(A + B) = P(A) + P(B)$$

Now P(A) = $\frac{1}{4}$ and P(B) = $\frac{3}{8}$; hence P(R) = P(A + B) = $\frac{1}{4} + \frac{3}{8} = \frac{5}{8}$

In general if A, B, C, ... N are mutually exclusive events then

$$P(A + B + C + \ldots + N) = P(A) + P(B) + P(C) + \ldots + P(N)$$

Dependence and independence

It is necessary to distinguish between dependent and independent events. When two or more events take place, these events are said to be *independent* when the outcome of one event makes no difference to the outcome of the other event(s).

In answering examination questions involving two or more events it is essential to determine whether the events are dependent or independent. In the case of problems which depend on drawing items out of a bag, box or other container, it is essential to see whether the question states or infers 'with replacement' or 'without replacement'. Suppose we have a bag which contains 5 red balls and 5 white balls.

We draw out a ball out of the bag and it is red; what is the probability that if we randomly select another ball from the bag it will be a white ball? If we *replace* the red ball in the bag before the second draw, then the probability that the ball is white is $\frac{5}{10}$ and is *independent* of the result of the first draw. If *we do not replace* the red ball in the bag there are now 4 red balls and 5 white balls in the bag and the probability that the ball is white is $\frac{5}{9}$ and is *dependent* on the result of the first draw.

It is not unknown for examination questions to be set where it is not completely clear whether the question is 'with replacement' or 'without replacement'. In this situation it is best to point out that the question is not clear and then to state the assumption on which you have solved the problem.

Multiplication law of probabilities

If A, B, C, ... N are independent events and the respective probabilities that these events occur are P(A), P(B), P(C) ... P(N), then the probability that *all* the events A, B, C, ... N occur P(A, B, C, ... N) is obtained by the multiplication of the probabilities.

i.e. $P(A, B, C, \ldots N) = P(A) \times P(B) \times P(C) \times \ldots \times P(N)$

Worked Example No. 8.1

A bag contains 4 red balls and 16 white balls. Two balls are drawn with replacement. What is the probability that both are red?

Let A be the event 'first ball red'; $P(A) = \frac{4}{20} = 0.2$

Let B be the event 'second ball red'; $P(B) = \frac{4}{20} = 0.2$.

$P(\text{Both red}) = P(A, B) = P(A) \times P(B) = 0.2 \times 0.2 = 0.04$.

Worked Example No. 8.2

A part is made of two components, A and B. Component A is known to have a defective rate of 5%; component B has a defective rate of 3%. Find the probability that (i) both components are defective; (ii) one component is good and the other component is defective; (iii) the part is defective.

Let A be the event 'part A good', $\overline{A}$ be event 'part A defective'.
Thus $P(\overline{A}) = 0.05$ (5% = 0.05) and $P(A) = 0.95$
similarly $P(\overline{B}) = 0.03$, $P(B) = 0.97$

i Probability both defective $= P(\overline{A}\,\overline{B}) = P(\overline{A}) \times P(\overline{B}) = 0.05 \times 0.03 = 0.001\,5$.

ii Probability one good and one defective = probability of A good and B defective or B good and A defective. These are mutually exclusive events, thus required probability $= P(A\,\overline{B}) + P(B\,\overline{A}) = P(A) \times P(\overline{B}) + P(B) \times P(\overline{A}) = (0.95 \times 0.03) + (0.97 \times 0.05) = 0.028\,5 + 0.048\,5 = 0.077$.

iii Probability that part is defective = sum of the results of parts (i) and (ii) $= 0.001\,5 + 0.077 = 0.078\,5$

If you were not asked to do parts (i) and (ii) the easiest way to solve part (iii) is to note that:

Probability of a good part + Probability of a defective part = 1

Thus probability of a defective part = 1 — probability of a good part

Now probability of a good part = P(A B) = P(A) × P(B) = 0.95 × 0.97 = 0.921 5
Thus probability that part is defective = 1 − 0.921 5 = 0.078 5

Exercise 8.2

a An electrical component is made up of two independent circuits, A and B, and is designed so that the component functions if at least one of the circuits is working. Experience shows that circuit A has a 5% chance of failing before it is one year old and circuit B has a 4% chance of failing before it is one year old. Find the probability that, at the end of one year, (i) both circuits are working, (ii) the component is working.

b A carton of 20 components contains 4 defective items. Three components are selected at random from the carton, with replacement. Find the probability that (i) all three components are good (ii) two components are good and one is defective.

Worked Example No. 8.3

Two true six-sided dice are rolled. (a) What is the probability of a total score of (i) 12, (ii) 10? (b) What is the probability that at least one dice shows a 6?

a To score 12 we must roll 2 sixes
$P(12) = P(6) \times P(6) = \frac{1}{6} \times \frac{1}{6} = \frac{1}{36}$
To score 10, possibilities are (4, 6) (6, 4) (5, 5)
$P(4, 6) = P(4) \times P(6) = \frac{1}{6} \times \frac{1}{6} = \frac{1}{36}$ similarly $P(6, 4) = \frac{1}{36}$ and $P(5, 5) = \frac{1}{36}$
$P(10) = P(4, 6) + P(6, 4) + P(5, 5) = \frac{1}{36} + \frac{1}{36} + \frac{1}{36} = \frac{3}{36} = \frac{1}{12}$

b The easiest way to solve probability questions which ask for the probability of 'at least one' is usually to note that:
Probability of 'at least one' + probability of 'none' = 1
$P(\text{at least one six}) = 1 - P(\text{no sixes})$
$= 1 - P(\overline{6}, \overline{6}) = 1 - P(\overline{6}) \times P(\overline{6})$
$= 1 - \frac{5}{6} \times \frac{5}{6} = 1 - \frac{25}{36} = \frac{11}{36}$

Exercise 8.3

Two fair ten-sided dice, each marked with numbers 1–10, are thrown simultaneously. Find the probability that

i the total score is 4 or less,
ii at least one dice shows a 2,
iii either (i) or (ii) is satisfied.

Conditional probability

When events are not independent we have to use conditional probability, since the result of the second event is conditional on the result of the first event.

If event A occurs and then event B follows, the conditional probability that event B occurs given that event A has occurred is expressed in symbols as:

$P(B/A)$

Worked Example No. 8.4

A bag contains 4 red balls and 16 white balls. Two balls are drawn without replacement, what is the probability that both are red?

(This is the same as Worked Example No. 8.1 but in this case balls are drawn without replacement.)

Let A be the event 'first ball red'; $P(A) = \frac{4}{20}$.

Let B be the event 'second ball red'; there are now 19 balls in the bag, of which 3 are red (since a red ball was taken at the first draw); thus $P(B/A) = \frac{3}{19}$

Hence $P(2 \text{ red balls}) = P(A) \times P(B/A) = \frac{4}{20} \times \frac{3}{19} = \frac{12}{380} = \frac{3}{95}$

Exercise 8.4

Find the probabilities of Exercise 8.2(b) in the case of no replacement.

Tree diagrams

For some probability problems there are two methods of solution. One method is to use the methods we have already met; the second method is to use a tree diagram. Many students find a tree diagram easier.

The following example will be solved by both methods.

100 members of a social club were questioned on whether they smoked or not. There were equal numbers of males and females. Half the males smoked and 40% of the females. Find the probability that if one of the members were chosen at random, (i) the person was a smoker; (ii) the person was female; (iii) the person was a female that smoked; (iv) the person was female given that the person selected was a smoker.

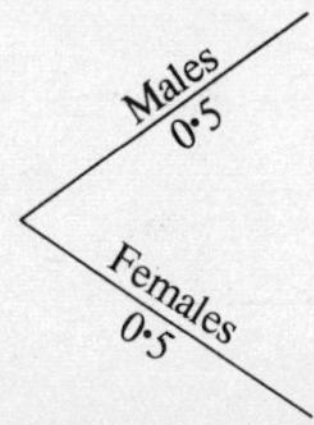

Figure 8.1

The first part of the question states that males and females are equally divided. The first part of the tree diagram is drawn in Figure 8.1.
The probabilities are entered under the corresponding branches. We now add the information about smokers and non-smokers as in Figure 8.2.

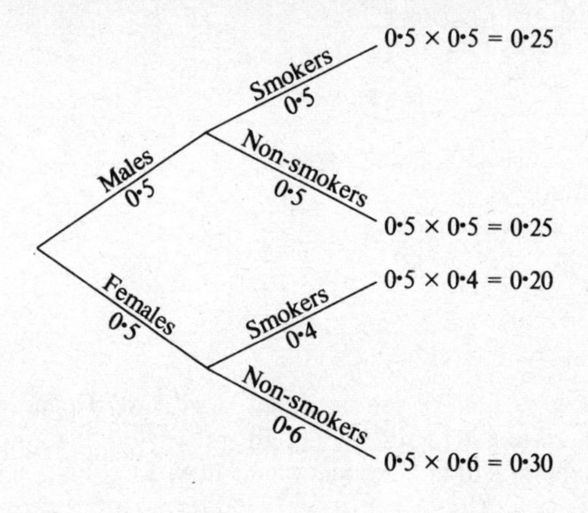

Figure 8.2

We now multiply the probabilities along the branches. (These have been added to Figure 8.2.) As a check on arithmetical accuracy, note that 0.25 + 0.25 + 0.20 + 0.30 must add to 1, since the tree covers all possibilities.
We can now read off the required probabilities.

i P(person a smoker). We identify the branches which involve smokers and add the probabilities: 0.25 + 0.20 = 0.45
ii P(person female). This is lower half of tree thus probability = 0.5
iii P(person was female that smoked). We identify the branch that is female-smoker and read off probability = 0.20
iv P(person female given a smoker selected). This is given by the probability that person was a female smoker divided by the probability that person was a smoker: $\dfrac{0.20}{0.25 + 0.20} = \dfrac{0.20}{0.45} = \dfrac{4}{9}$

N.B. Part (iv) is often referred to as *Bayesian methods*.
To solve using symbols it is helpful to express the information in the following table.

	Male	*Female*	*Total*
Smoker	25	20	45
Non-smoker	25	30	55
Total	50	50	100

Let A be event smoker, Let B be event female

i $\text{P(person a smoker)} = \text{P(A)} = \dfrac{\text{Number of smokers}}{\text{Number of members}} = \dfrac{45}{100} = 0.45$

ii $P(\text{person female}) = P(B) = \dfrac{\text{Number of females}}{\text{Number of members}} = \dfrac{50}{100} = 0.5$

iii P(person was female that smoked)

$$= P(A, B)$$

$$= \frac{\text{Number of females that smoked}}{\text{Number of members}}$$

$$= \frac{20}{100} = 0.20$$

iv P(person female given a smoker selected)

$$= P(B/A)$$

$$= \frac{\text{Number of female smokers}}{\text{Number of smokers}}$$

$$= \frac{20}{45} = \frac{4}{9}$$

From the above results we can note that if we have dependent events, then P(A, B) is *not* equal to P(A) × P(B). P(A, B) = 0.20 while P(A) × P(B) = 0.45 × 0.5 = 0.225. In the case of independent events P(A, B) = P(A) × P(B). If events are independent, then P(B/A) = P(B) since it does not affect the probability of B whether A occurs or not. An examination of P(B/A) and P(B) is a basis of finding out whether A and B are dependent. In this example P(B/A) = $\frac{4}{9}$, P(B) = 0.5. As P(B/A) is not equal to P(B), this shows that events A and B are dependent.

Worked Example No. 8.5

A manufacturer has launched a new consumer durable product. The marketing manager thinks that the chances of success or failure are even. He knows from past experience of other products that if a product is going to be a success, the probability that the rate of return of guarantees during the first month will be high is 0.8; on the other hand, if the product is going to be a failure, the probability that the rate of return of guarantees in the first month will be high is 0.3.

If the rate of return of guarantees is high in the first month, how will the Marketing Manager's assessment of the success of his product change?

This is an example where we are using additional information (high return of guarantees) in the calculation of probabilities. This is usually called *Bayesian methods*. (See Figure 8.3)

The probability of obtaining high returns is therefore 0.4 + 0.15 = 0.55. If we have high returns, then the manager's assessment of the success of the product is given by

$$P(\text{Success given high returns}) = \frac{P(\text{Success and high returns})}{P(\text{high returns})}$$

$$= \frac{0.4}{0.4 + 0.15} = \frac{0.4}{0.55} = 0.727$$

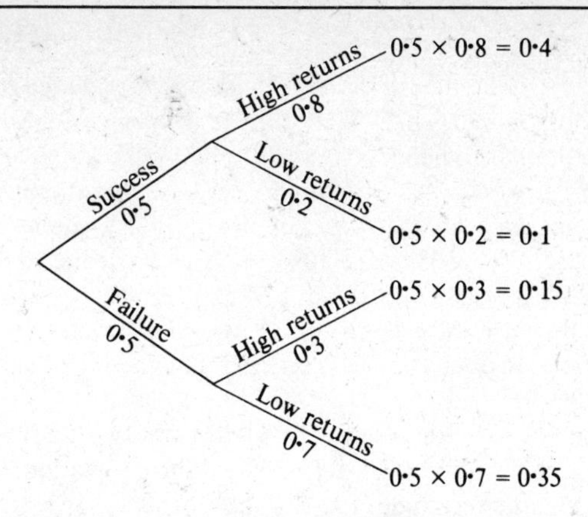

Figure 8.3

Likewise, if the return of guarantees were low, then the probability of failure given low returns

$$= \frac{\text{P(failure and low returns)}}{\text{P(low returns)}} = \frac{0.35}{0.35 + 0.1} = \frac{0.35}{0.45} = 0.778$$

Exercise 8.5

Each day a machine is set up. It is only possible to say whether it has been correctly set up after examination of the output of the machine. Experience shows that 6 days out of 10 the machine is correctly set up. 3 days out of 10 the machine needs minor adjustment and 1 day out of 10 the machine needs major adjustment. If the machine is correctly set up, 95% of the output is of good quality. If the machine requires minor adjustment, 80% of the output is of good quality. If the machine requires major adjustment, 30% of the output is of good quality.

i On a given day the first part is of poor quality. What is the probability that the machine requires major adjustment?

ii On another day the first 3 parts are of good quality. What is the probability that the machine is correctly set up?

(B.A.S.S. (P/T) 1977)

Addition laws of probability when events are not mutually exclusive

When events (A, B) are mutually exclusive, then the probability that one or other event A or B will occur is P(A + B) = P(A) + P(B). We need to see how this equation is modified when events are not mutually exclusive, i.e. events A and B can occur at the same time.

Consider the following example: suppose that a survey of a particular town shows that 40% of the population watch ITV, and 20% read the *Mirror*. Then if an advertiser placed an advert on ITV and in the *Mirror* it would be incorrect to assume that his potential audience is 40 + 20 = 60% of the population, since many who read the *Mirror* would also watch ITV. Suppose that the survey reveals that 10% of the population read the *Mirror* and watch ITV. Its potential audience would be 40 + 20 − 10 = 50%. The 10% needs to be deducted to avoid double counting. This analysis may be shown in a Venn diagram.

The 30% watching ITV and not reading the *Mirror* is obtained by deducting 10% from 40%. Likewise we obtain the 10% reading the *Mirror* and not watching ITV, 20% − 10% = 10%.

The area in the rectangle represents the whole population of the town. As the potential audience of the advertisement is 50% of the population, the area outside the circles but inside the rectangle (100% − 50% = 50%) represents the audience who do not see the advertisement.

In symbols this is equivalent to $P(A + B) = P(A) + P(B) - P(A\ B)$

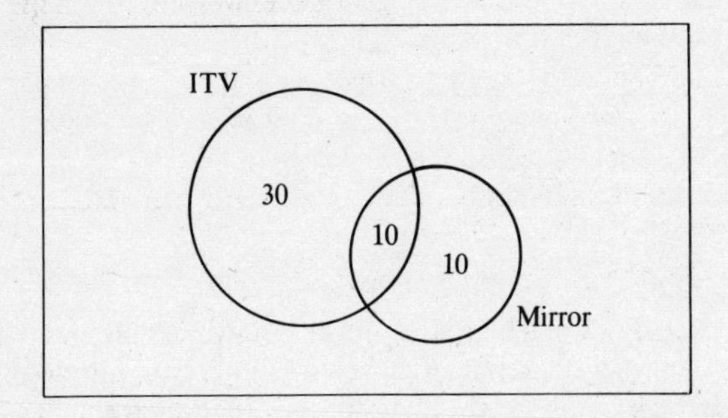

Figure 8.4

Suppose now that the advertiser decides additionally to advertise in the *Sun*. What is now the audience if the survey shows readership of the *Sun* is 20% of the population. 12% watch ITV and read the *Sun*, 5% read the *Sun* and the *Mirror*, and 3% watch ITV and read the *Sun* and the *Mirror*. What will be his total audience?

Use a Venn diagram. It is necessary to work out the percentages in each area.

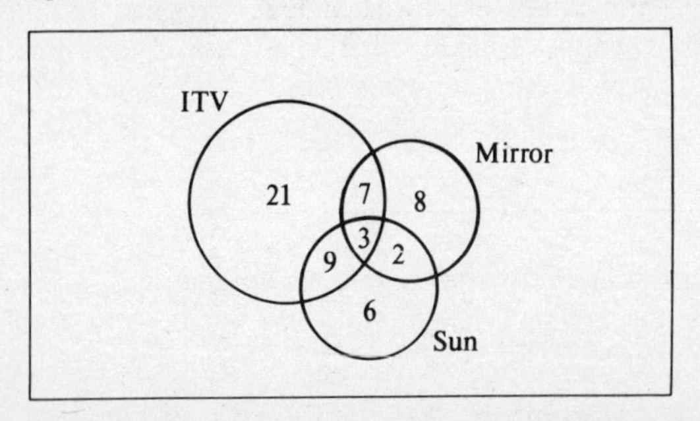

Figure 8.5

If 12% watch ITV and read the *Sun*, and 3% watch ITV and read the *Sun* and *Mirror*, then 12–3 = 9% watch ITV and read the *Sun* but do not read the *Mirror*. Similarly for the other intersections. We can then evaluate those who watch ITV but do not read the *Sun* or *Mirror* as 40–7–3–9 = 21%; similarly for *Sun* and *Mirror*. Thus potential audience is 21 + 6 + 8 + 7 + 9 + 2 + 3 = 56%

There is a formula

$$P(A + B + C) = P(A) + P(B) + P(C) - P(AB) - P(AC) - P(BC) + P(ABC)$$
$$= 40 + 20 + 20 - 12 - 5 - 10 + 3 = 56$$

However, it is usually better to use the Venn diagram, since other information is also easily available i.e. the percentage that have one opportunity only to see the advert (21 + 8 + 6 = 35%), and the percentage who have two opportunities to see the advert (7 + 9 + 2 = 18%).

When solving problems of this type using a Venn diagram it is essential to start at the three-overlap point and work outwards. In this example enter the 3 first and then calculate the double overlap areas; as 12% watch ITV and read the *Sun*, subtract 3 from 12 = 9 and then enter 9 on the diagram. A common error is to enter 12 instead of 9.

Exercise 8.6

100 cars were given the Ministry of Transport test. 34 had defective brakes, 30 had defective steering, 26 had defective tyres. 14 had both brakes and steering defective, 10 had tyres and steering defective, 13 had brakes and tyres defective, and 6 had brakes and tyres and steering defective.

Find the percentage of cars that had just one defect and the percentage of cars that passed the test with no defects.

(B.A.S.S. (P/T) 1978)

Revision Exercises

8.7 A box contains 16 components. This box was dropped in transit and six components were defective, but not visibly defective.

a What is the probability that if **i** 1, **ii** 2, **iii** 3 components are selected at random they are all good?

b Components are taken from the box and tested until a good item is found. What is the probability that the first good component was **i** 1st, **ii** 2nd, **iii** 4th component tested?

8.8 A salesman for a company sells two products, A and B. During a morning he makes three calls on customers. If the chance that he makes a sale of product A is 1 in 3 and the chance that he makes a sale of product B is 1 in 4, and assuming that the sale of product A is independent of the sale of product B, calculate the probability that the salesman:

i will sell both product A and product B at the first call;
ii will sell *one* product at the first call;
iii will make no sales of product A in the morning;
iv will make at least one sale of product B in the morning.

(I.O.S. 1975)

8.9 **a** There are 30 students taking a course in which they must include one optional subject. Twelve students study Politics, and 18 study Sociology as their option. Two students are selected at random without replacement from the 30.

Find the probability that **i** both study Politics and **ii** one studies Politics and the other Sociology.

b On another course students may study one, two or three optional subjects. Sixteen students study Sociology, 21 study Politics, 20 study History, 7 study Politics and Sociology, 8 study Politics and History, 5 study Sociology and History and 3 study Sociology, Politics and History.

i Calculate the total number of students on the course.
ii Find the probability that a student selected at random studies one optional subject only.

(I.O.S. 1976)

8.10 A part is made up of 3 components which have the following probabilities of working satisfactorily: 0.96, 0.98, 0.99. The part is designed so that it still functions with only one component working. Calculate the probability that **a** the part fails, **b** has only one component working, **c** has at least one component working.

8.11 A building contractor submits 3 tenders for jobs with 3 different firms. He considers that his chances of getting the contracts are $\frac{1}{2}$, $\frac{1}{3}$, $\frac{1}{4}$ respectively. What is the probability that he will obtain **a** one and only one contract, **b** no contracts?

8.12 Let X and Y denote the scores obtained in throwing two fair six-sided dice. Find the probability that

a a previously specified dice yields a score of at least 4,
b at least one of the dice yields a score of at least 4,
c the score on each dice is at least 4,
d the sum of the two scores is at least 4,
e the numerical value (ignoring sign) of the difference between the two scores is

i exactly 4,
ii at least 4.

(I.O.S. 1965)

8.13 A Government Committee consists of 16 men and 12 women. Nine of the men and 8 of the women are members of the blue party. The remainder are members of the red party. Three members are selected at random, without replacement, from the 28 committee members. What is the probability that the selected members are:

i all men,
ii all members of the blue party,
iii all female members of the red party,
iv such that the red party members are in the majority?

(I.O.S. 1977)

8.14 Our local darts team is made up of seven men and five women. Each year we choose a captain and a secretary. We have a two-stage selection procedure:

Stage 1 The names of the twelve members of the team are put in a hat and two are drawn at random.

Stage 2 The two people chosen in stage 1 have a darts match. The winner is chosen as captain, the loser as secretary.

The men are virtually the same standard, and similarly the women are the same standard. But if a man plays a match against a woman, the man has twice as much chance of winning as the woman.

What is the probability that:

i both the captain and secretary are male,
ii both the captain and secretary are female,
iii the captain and secretary are not the same sex,
iv the captain is male and the secretary female,
v the captain is female and the secretary male?

(B.A.B.S. 1976)

8.15 a A machine shop has machines of three different types, with details as given below:

Type	*A*	*B*	*C*
Percentage of total no. of machines	20	40	40
Percentage of defective items produced	6	3	1

i What is the overall percentage of defective items?
ii An item is examined and found to be defective. What is the probability that it was produced by a Type A machine?

b A porter in an office block has made 10 master keys which together will open every door in the block. The doors are allocated at random to the master keys. A door can be opened by only one of the keys. If 30% of the doors at random are usually left unlocked, what is the probability that the porter will be able to open the door to a specific office if he selects 5 master keys at random?

(B.A.B.S. 1976)

8.16 In a country there are three Sunday newspapers; the *Echo*, *Advertiser* and *News*. A survey is conducted and it is found that 32% read the *Echo*; 38% read the *Advertiser*; 40% read the *News*; 7% read the *Echo* and the *Advertiser*; 9% read the *Advertiser* and the *News*; 8% read the *Echo* and the *News*; 3% read all three newspapers.

Find **a** the percentage that read *either* the *Echo* or the *News* *or* both *Echo* and *News*;

b the percentage that read at least one Sunday newspaper;

c the percentage that read only the *Advertiser*.

(I.O.S. 1976)

8.17 A multinational company recruits each year a large number of graduates. In the first two years the graduates spend a short period in a number of main departments of the company and then an assessment is made of their potential. 10% are sent on a full-time course in Management Studies for an MA Degree. 30% are allowed day release to follow a Diploma in Management Studies course. The remainder are not given time off but are refunded any expenses they incur on any relevant course which they attend.

An analysis of the subsequent promotion of graduate recruits shows that 60% of those with MA in Management Studies have reached the rank of senior executive by the age of 35. The corresponding percentages for those who have had day release and no time off were 20% and 5% respectively.

a Find the probability that a graduate recruit randomly selected will be a senior executive by the age of 35.

b Find the probability that a man who is a senior executive at the age of 35 has an MA in Management Studies from a full-time course.

(R.S.A. 1978)

9 The Normal, Binomial and Poisson Distributions

Normal distribution

The normal distribution is a very important distribution in the theory of statistics. In addition, it is the shape of the curve which fits certain types of data, e.g. heights of men, intelligence quotients (IQ) of children attending primary schools, errors of observations.

The normal distribution with mean 0 and standard deviation σ is shown in Figure 9.1.

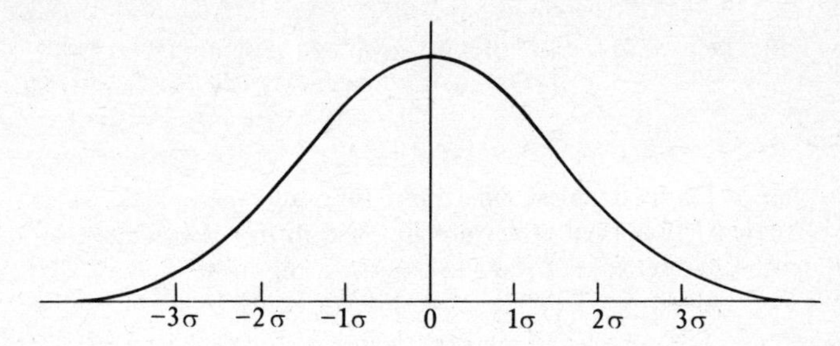

Figure 9.1

Data which are normally distributed have 68.3% of the observations within one standard deviation of the mean, 95.4% of the observations within two standard deviations of the mean and 99.7% of the observations within three standard deviations of the mean. Thus in the case of data which are normally distributed, nearly all the observations lie within the range of six standard deviations.

Normal distribution tables

The special tables (see page 303) are based on the standard normal distribution, i.e. with zero as the mean and a standard deviation of unity, $\sigma = 1$ (see Figure 9.2).

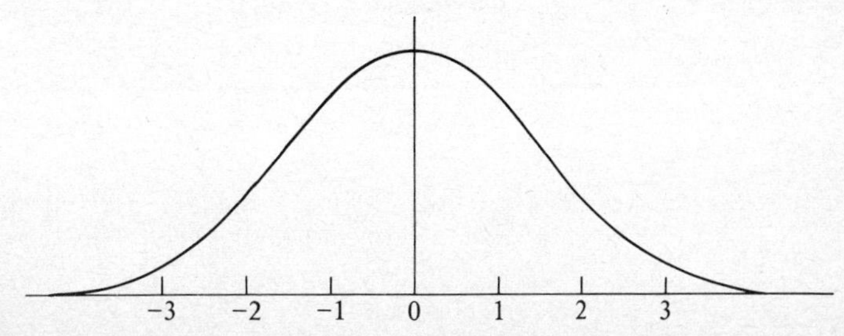

Figure 9.2

The tables give the area to the left of the ordinate x — the shaded area in Figure 9.3.

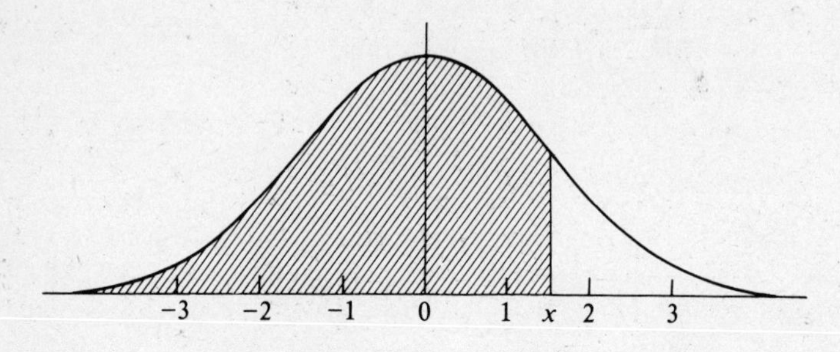

Figure 9.3

In the table the area to the left of x is shown by the symbol $\Phi(x)$.

Important note

There are various normal tables; some tables for example give the area to the *right* of the ordinate x. If normal tables are supplied in the examination room, it is essential to obtain a set so that you can practise using them.

Worked Example No. 9.1

Data are normally distributed with mean 0 and standard deviation 1. Find the area (i) to the left of the ordinate $x = 1.5$; (ii) to the right of the ordinate $x = 1.5$.

In Figure 9.3 the ordinate x has been drawn at $x = 1.5$. For part (i) we need to find the shaded area, and for part (ii) we need to find the unshaded area.

From the tables, when $x = 1.5$, $\Phi(x) = 0.933\ 2$. Thus the area to the left of $x(= 1.5)$ is 0.933 2.

The area under the normal curve is equal to 1; hence the area to the right of $x(= 1.5)$ is $1 - 0.933\ 2 = 0.066\ 8$.

Exercise 9.1

Data are normally distributed with mean 0 and standard deviation 1. Find the area (i) to the left of the ordinate x; (ii) to the right of the ordinate x, when x takes the following values: (a) 1.3, (b) 0.8, (c) 1.37.

Worked Example No. 9.2

If data are normally distributed with mean 0 and standard deviation 1, find the probability that an observation chosen at random lies between $x = 0.5$ and $x = 2$.

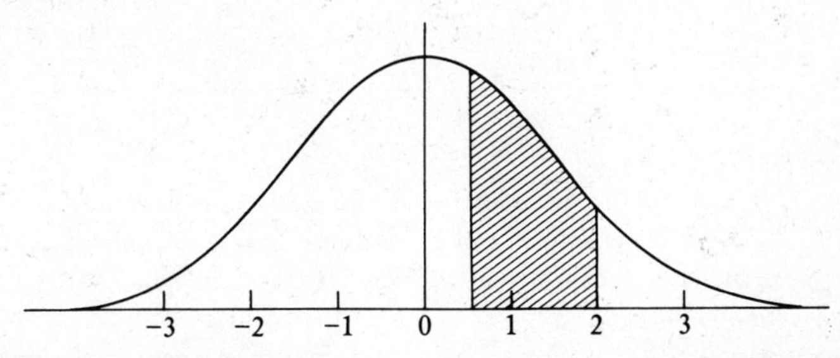

Figure 9.4

The required probability is equal to the shaded area of Figure 9.4

Area to the left of $x = 2$ is $\Phi(2) = 0.977\ 25$

Area to the left of $x = 0.5$ is $\Phi(0.5) = 0.691\ 5$

Required probability is $0.977\ 25 - 0.691\ 5 = 0.285\ 75$

Exercise 9.2

Data are distributed in the standard normal distribution. Find the probability that an observation lies between $x = 0.32$ and $x = 1.88$.

Worked Example No. 9.3

If data are normally distributed with mean 0 and standard deviation 1, find the area to the left of the ordinate $x = -1$ and hence find the area between $x = -1$ and $x = +1$.

The tables give values of x where x is positive. To find areas when x is negative we have to use the fact that the normal distribution is symmetrical about the mean.

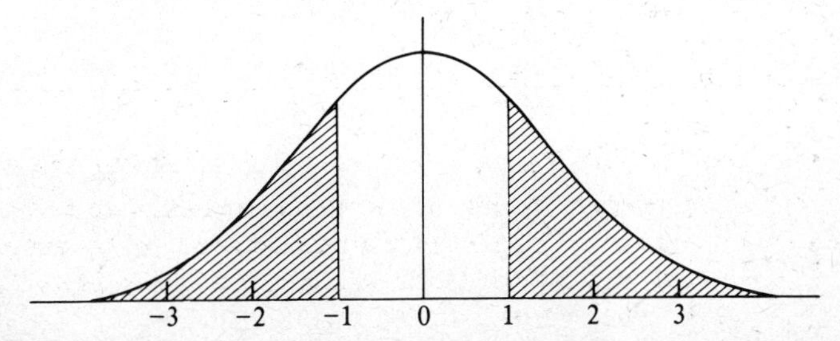

Figure 9.5

The area to the left of –1 is equal to the area to the right of +1.
The area to the right of +1 is equal to 1 minus the area to the left of +1,

i.e. $\Phi(-1) = 1 - \Phi(+1) = 1 - 0.841\,3 = 0.158\,7$

N.B. In general $\Phi(-x) = 1 - \Phi(x)$

Now area between +1 and –1 is $\Phi(+1) - \Phi(-1)$

$= 0.841\,3 - 0.158\,7 = 0.682\,6$

Exercise 9.3

If data are normally distributed with mean 0 and standard deviation 1, find the area to the left of the ordinate $x = -0.36$ and hence find the area between the ordinate $x = -0.36$ and $x = 1.62$.

Worked Example No. 9.4

If x is normally distributed with mean 0 and standard deviation 1, find the value of x if the area between $-x$ and $+x$ is (i) 0.95, (ii) 0.99.

We need to find the values of x such that the shaded area in Figure 9.6 is (i) 0.95, (ii) 0.99.

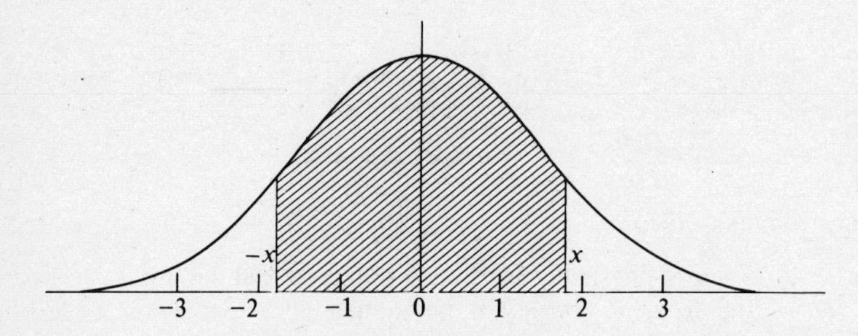

Figure 9.6

i If shaded area = 0.95, each tail (the unshaded area at each end) = 0.025, thus the area to the left of x is $0.95 + 0.025 = 0.975$

Thus $\Phi(x) = 0.975$

The tables show that the corresponding value of $x = 1.96$.

Thus 95% (=0.95) of observations lie between $x = +1.96$ and $x = -1.96$.

ii If shaded area = 0.99, each tail = 0.005, thus the area to the left of x is $0.99 + 0.005 = 0.995$

Thus $\Phi(x) = 0.995$

Hence $x = 2.58$

Thus 99% (=0.99) of observations lie between $x = +2.58$ and $x = -2.58$.

Exercise 9.4

If x is normally distributed with mean 0 and standard deviation 1, find the value of x if the area between $-x$ and $+x$ is 0.90.

To find probabilities when we have a normal distribution with mean other than zero and standard deviation other than unity

In these cases we transform the distribution to a normal distribution with mean 0 and standard deviation 1 so that we can use our tables.

We use $z = \dfrac{x - \mu}{\sigma}$ to transform to the standard normal distribution

where z = standardised normal variable
and μ = population mean
and σ = population standard deviation.

Worked Example No. 9.5

Data are normally distributed with mean $\mu = 2$ and standard deviation $\sigma = 5$. Find the probability that an observation chosen at random lies between $x = -5$ and $x = +10$.

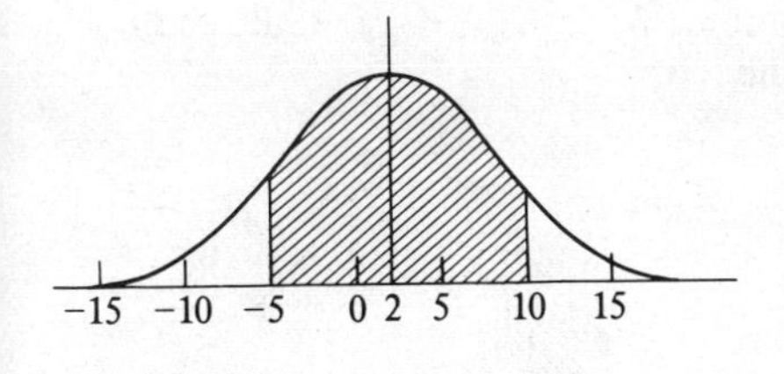

Figure 9.7

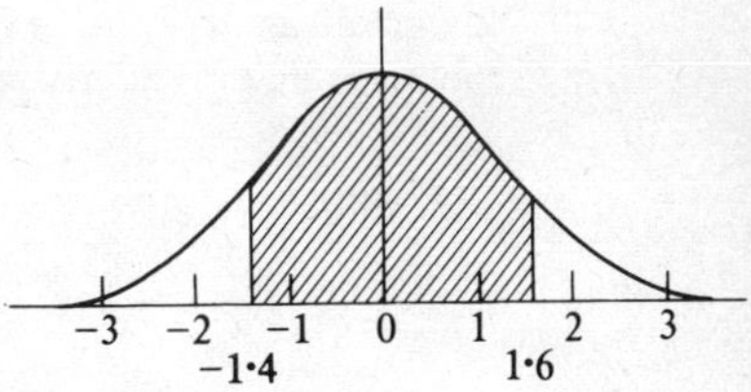

Figure 9.8

We need to find the area of the shaded region in Figure 9.7.

We use $z = \dfrac{x - \mu}{\sigma}$ to standardise.

$$\text{Now } \mu = 2, \quad \sigma = 5: z_1 = \frac{10-2}{5} = \frac{8}{5} = 1.6$$

$$z_2 = \frac{-5-2}{5} = \frac{-7}{5} = -1.4$$

Required area is also the shaded region in the standard normal distribution shown in Figure 9.8

This area is $\Phi(1.6) - \Phi(-1.4)$

$\Phi(1.6) = 0.945\,2$; $\Phi(-1.4) = 1 - \Phi(1.4) = 1 - 0.919\,2 = 0.080\,8$

Thus area is $0.945\,2 - 0.080\,8 = 0.864\,4$

Exercise 9.5

x has a normal distribution with mean 10 and standard deviation 4. Find the probability that an observation x selected at random (i) is less than 15, (ii) is greater than 12, (iii) is greater than 3, (iv) lies between 5 and 9.

Problems involving the normal distribution

Worked Examples Nos. 9.6 and 9.7 cover two of the standard questions on the normal distribution. Most questions on this topic involve standardising to the standard normal distribution. It is essential to draw a rough sketch of the normal curve and shade the required area. Remember when you draw your sketch that almost all the observations lie within 3 standard deviations of the mean. You should then check whether your answer is approximately correct.

Worked Example No. 9.6

Flour is packed in bags with a nominal weight of 1.5 kg. The mean weight of bags of flour is found to be 1.510 kg with a standard deviation of 8 g. Assuming that bags of flour are normally distributed, find the percentage of bags with a weight less than 1.5 kg. If not more than one bag in twenty is to contain less than 1.5 kg, what value must the mean have to achieve this?

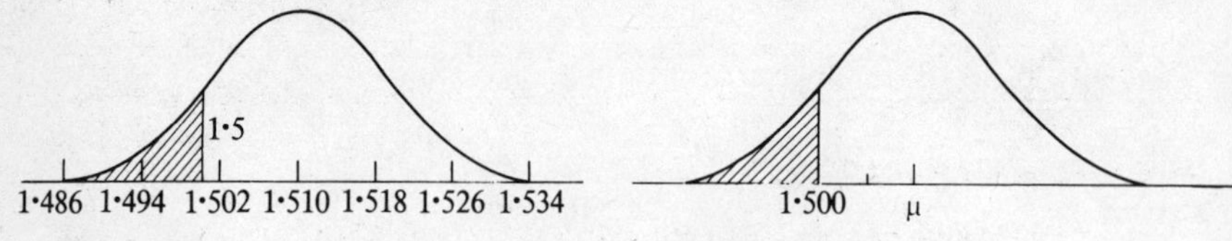

Figure 9.9

Figure 9.10

To find the percentage of bags weighing less than 1.5 kg we need to find the shaded area in Figure 9.9.

$$z = \frac{x - \mu}{\sigma} = \frac{1.500 - 1.510}{0.008} = \frac{-0.010}{0.008} = -1.25$$

Area is $\Phi(z) = \Phi(-1.25) = 1 - \Phi(1.25) = 1 - 0.894\ 4 = 0.105\ 6$

Percentage under-weight is $0.105\ 6 \times 100 = 10.56\%$

One in twenty is equal to 5% or 0.05. We need to find the position of the mean μ so that the shaded area in Figure 9.10 $= 0.05$ and the unshaded area $= 0.95$. From the normal tables we need to find x so that $\Phi(x) = 0.95$, i.e. $x = 1.645$. The mean μ needs to exceed 1.5 by

1.645 σ i.e. $\mu = 1.5 + 1.645\sigma$

now $\sigma = 8$ g $= 0.008$ kg

Thus $\mu = 1.5 + 1.645 \times 0.008 = 1.513$ kg

Hence the average weight of bags of flour must be at least 1.513 kg to ensure that not more than one bag in twenty is less than 1.5 kg.

N.B. If not more than one bag in a hundred is to contain less than 1.5 kg, the corresponding result is $\mu = 1.5 + 2.33\sigma$; $\mu = 1.5 + 2.33 \times 0.008 = 1.519$ kg

---

Exercise 9.6

The following data show the daily sales of a commodity in a shop during a period of 100 days.

Number sold	*Number of days*
Under 100	3
100 and under 120	13
120 and under 140	20
140 and under 160	35
160 and under 180	16
180 and under 200	10
200 and over	3

(a) Calculate the mean and standard deviation of daily sales.

(b) Assuming that the distribution of sales is normal and that stock cannot be replaced until the following day, how many items should be held in stock if the shop is not to run out of stock more frequently than one day in twenty?

(R.S.A. 1977)

Worked Example No. 9.7

A component has a length which is normally distributed with mean 10 cm and standard deviation 0.04 cm. The cost of production is 5 pence per component. A component of length between 9.94 and 10.06 cm is acceptable and can be sold for 10 pence. Components outside these dimensions can be sold for scrap for 2 pence. Find the profit per 1 000 components.

We need to find the shaded area in Figure 9.11

$$z_1 = \frac{x - \mu}{\sigma} = \frac{10.06 - 10}{0.04} = 1.5 \quad z_2 = \frac{x - \mu}{\sigma} = \frac{9.94 - 10}{0.04} = -1.5$$

Shaded area $= \Phi(1.5) - \Phi(-1.5) = 0.933\,2 - (1 - 0.933\,2) = 0.866\,4$

Unshaded area $= 1 - 0.866\,4 = 0.\,133\,6$

For each 1 000 components, 1 000 $\times$ 0.866 4 $=$ 866 can be sold for 10 pence and 1 000 $\times$ 0. 133 6 $=$ 134 can be sold for scrap for 2 pence.

Thus revenue per 1 000 components is (866 × 10) + (134 × 2) = 8 928 pence. The cost per 1 000 components is 1 000 × 5 = 5 000 pence. Hence profit per 1 000 components is 8 928 - 5 000 = 3 928 pence = £39.28.

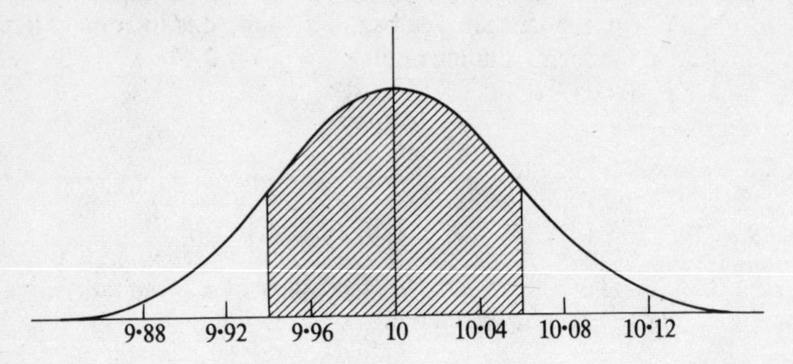

Figure 9.11

Binomial and Poisson distributions

Important note

Many examination syllabuses include the normal distribution but *exclude* the binomial and poisson distributions. It is necessary, therefore, to consult the syllabus and/or past examination papers to see whether the final part of this chapter needs to be studied. Exercises on these distributions are included at the end of the chapter.

Binomial distribution

The binomial distribution is used when we have a series of n independent trials and the probability of success at each trial is constant and is equal to p, and the probability of failure is equal to $1 - p = q$.

Consider two examples of the Binomial distribution. (1) We toss a coin 10 times. What are the probabilities that we obtain 0, 1, 2, 3, 4, ... 10 heads? In this situation, n = number of trials = 10, and the probability of a head in a single trial = p = 0.5. (We are assuming that a head and a tail are equally likely.) (2) Mass-produced components are known to have a defective rate of 5%. A random sample of 5 items is taken from the production line. What is the probability that the sample contains 0, 1, 2, 3, 4, or 5 defective items? In this case $n = 5$, p = 0.05 (5% = 0.05).

Factorial notation

With the binomial and poisson distributions we use factorials. 4! (we say factorial four) is equal to $4 \times 3 \times 2 \times 1 = 24$; $5! = 5 \times 4 \times 3 \times 2 \times 1 = 120$; $6! = 6 \times 5 \times 4 \times 3 \times 2 \times 1 = 720$ and so on.
N.B. $1! = 1$, $0! = 1$

It is possible to show that the probability of exactly r successes in n trials is given by the formula

$$P(r) = \frac{n!}{(n-r)!\,r!} p^r q^{(n-r)}$$

where r takes the values 0, 1, 2, 3, ... n

On substituting r = 0, 1, 2, 3 ... n in this formula, we find on simplification that

$$P(0) = q^n; \quad P(1) = npq^{n-1}; \quad P(2) = \frac{n(n-1)}{2} p^2 q^{n-2};$$

$$P(3) = \frac{n(n-1)(n-2)}{3 \times 2} p^3 q^{n-3};$$

$$P(4) = \frac{n(n-1)(n-2)(n-3)}{4 \times 3 \times 2} p^4 q^{n-4}; \ldots \; P(n) = p^n$$

Let us consider again the example on defectives. There were samples of 5, so n = 5. The probabilities of a defective was 0.05 so p = 0.05 and q = (1–p) = 1 – 0.05 = 0.95. Now the probability of no defectives is the first term in the expression above.

$$P(0) = q^n = (0.95)^5 = 0.773\,8$$

The probability of 1 defective is the second term in the expression above.

$$P(1) = npq^{n-1} = 5 \times 0.05 \times (0.95)^4 = 0.203\,6$$

Similarly

$$P(2) = \frac{n(n-1)}{2} p^2 q^{n-2} = \frac{5 \times 4}{2} \times (0.05)^2 \times (0.95)^3 = .021\,4$$

$$P(3) = \frac{n(n-1)(n-2)}{3 \times 2} p^3 q^{n-3} = \frac{5 \times 4 \times 3}{3 \times 2} \times (0.05)^3 \times (0.95)^2 = 0.001\,1$$

$$P(4) = \frac{n(n-1)(n-2)(n-3)}{4 \times 3 \times 2} p^4 q^{n-4} = \frac{5 \times 4 \times 3 \times 2}{4 \times 3 \times 2} \times (0.05)^4 \times (0.95)$$

$$= 0.000\,03$$

$$P(5) = p^n = (0.05)^5 = 0.000\,000\,3$$

Special tables are available which enable us to find the probabilities without so much tedious arithmetic. Unfortunately, for most examinations these special tables are not provided. The tables show the probability of obtaining *r or more* successes. Thus in our example p = 0.05, n = 5. The tables give

(n = 5, p = 0.05)

r	
r = 0	1.0000
1	0.2262
2	0.0226
3	0.0012
4	—

There is no entry for r = 4, since the probability is less than 0.000 05.

1.0000 is the probability of 0 or more successes
0.2262 is the probability of 1 or more successes
0.0226 is the probability of 2 or more successes
0.0012 is the probability of 3 or more successes

Hence, 1.0000 – 0.2262 = 0.773 is the probability of 0 success
0.2262 – 0.0226 = 0.2036 is the probability of 1 success
0.0226 – 0.0012 = 0.0214 is the probability of 2 successes

N.B. The tables give the binomial probabilities for only a limited range of values of n and p.

The calculation of the probabilities can become particularly tedious if n is large.

Fortunately, in some situations, i.e. if $np \geqslant 5$, the normal distribution can be used to give a good approximation to the binomial distribution.

Normal approximation to the binomial distribution

Consider again the example of tossing a coin 10 times. The probabilities of 8 or more heads is given by P(8) + P(9) + P(10). Here n = 10, p = 0.5. We find P(8) = 0.0440, P(9) = 0.009 7, P(10) = 0.001 0; thus the exact result is that the probability of 8 or more heads is 0.044 0 + 0.009 7 + 0.001 0 = 0.054 7.

We may show that the mean of a binomial distribution is equal to np and the standard deviation is equal to $\sqrt{np(1-p)}$. The mean is therefore $10 \times 0.5 = 5$, the standard deviation is $\sqrt{10 \times 0.5 \times (1-0.5)} = \sqrt{2.5} = 1.581$, i.e. $\mu = 5$, $\sigma = 1.581$. In figure 9.12 we are using the normal curve to find approximately the area of the three rectangles. It will be noted that to find the required area we have to measure from 7.5 upwards.

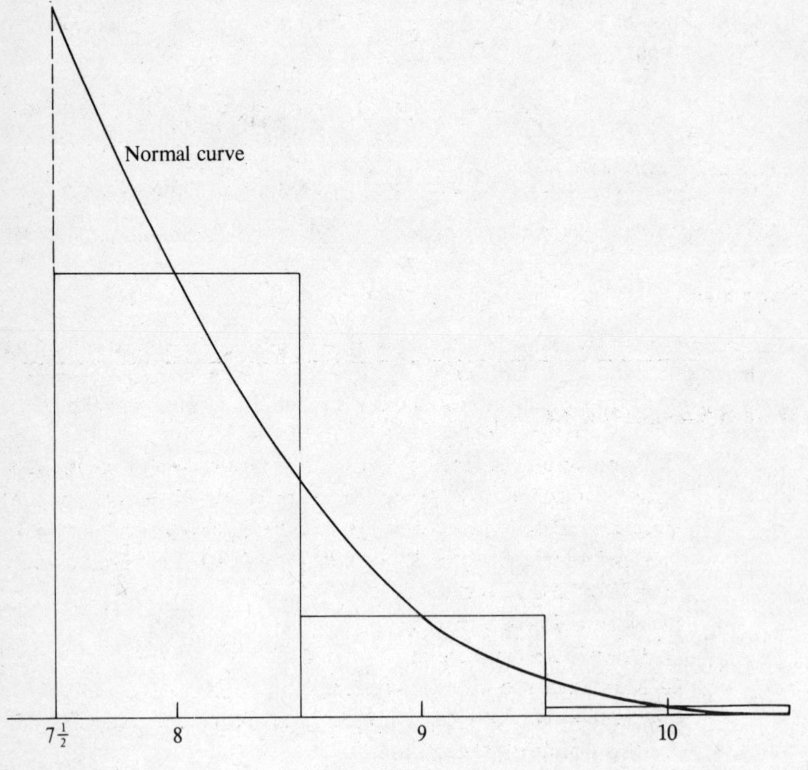

Figure 9.12

We need to find the probability that an observation is greater than 7.5 when drawn from a normal population with mean $\mu = 5$ and $\sigma = 1.581$.

$$z = \frac{x - \mu}{\sigma}$$

$$= \frac{7.5 - 5}{1.581}$$

$$= \frac{2.5}{1.581} = 1.581$$

$$\text{Area} = 1 - \Phi(z)$$
$$= 1 - \Phi(1.581)$$
$$= 1 - 0.943\,0$$
$$= 0.057\,0$$

The exact result was 0.054 7; the normal approximation is therefore quite close.

In general, the larger the value of n, the closer the normal approximation to the exact result. In this example $n = 10$, which is a relatively small value of n for the use of the normal approximation.

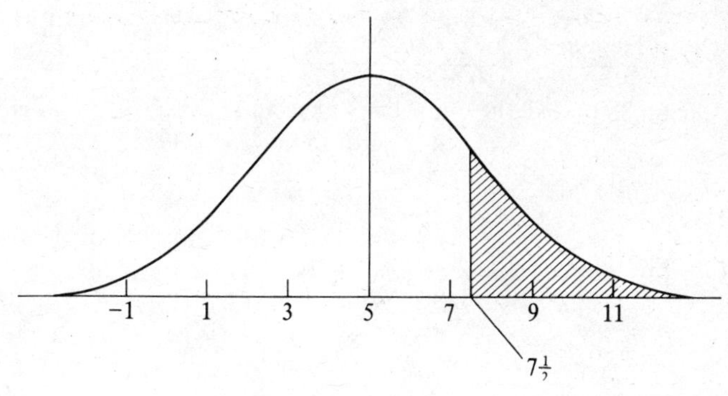

Figure 9.13

Worked Example No. 9.8

A manufacturer guarantees that he will repair, free of charge, any washing machine that proves to be defective within one year of purchase. Past experience suggests 10% of machines have to be repaired under guarantee. What is the probability that from a random sample of 20 machines

i None will need repairing;
ii One will need repairing;
iii More than two will need repairing?

The average cost of a repair is £25. Calculate the expected repair bill for this sample of 20 machines.

We have $n = 20$, $p = 0.1$, $q = 0.9$

i $P(0) = q^n = (0.9)^{20} = 0.1216$

ii $P(1) = npq^{n-1} = 20 \times 0.1 \times (0.9)^{19} = 0.2702$

iii $P(r > 2) = 1 - P(0) - P(1) - P(2)$

$$P(2) = \frac{n(n-1)}{2}p^2q^{n-2} = \frac{20 \times 19}{2} \times (0.1)^2 \times (0.9)^{18} = 0.2852$$

$\therefore$ P($r > 2$) = 1 – 0.1216 – 0.2702 – 0.2852 = 0.3230
The mean number of repairs = np = 20 × 0.1 = 2
Thus expected repair bill = 2 × £25 = £50

Worked Example No. 9.9

A examination consists of 100 multiple-choice questions. For each question there are 4 answers, of which one is correct. A candidate enters the examination who knows nothing about the subject and he choses answers randomly. Find the probability that he obtained (i) less than 20 marks; (ii) 35 or more correct answers; (iii) exactly 25 correct answers.

Here n = 100, p = 0.25; thus μ = np = 100 × 0.25 = 25 σ = $\sqrt{np(1-p)} = \sqrt{100 \times 0.25 \times 0.75} = 4.33$

(i) We need the probability that there were 19 or fewer correct answers. We need the area to the left of 19.5.

$$z = \frac{x-\mu}{\sigma} = \frac{19.5-25}{4.33} = -1.27$$

Area = Φ(–1.27) = 1 – Φ(1.27) = 1 – 0.898 = 0.102.

(ii) We need the area to the right of 34.5.

$$z = \frac{x-\mu}{\sigma} = \frac{34.5-25}{4.33} = 2.194$$

Area = 1 – Φ(2.194) = 1 – 0.986 = 0.014

(iii) We need the area between 24.5 and 25.5.

$$z_1 = \frac{x-\mu}{\sigma} = \frac{24.5-25}{4.33} = -0.115$$

$$z_2 = \frac{25.5-25}{4.33} = 0.115$$

Area = Φ(0.115) – Φ(–0.115) = 0.546 – (1 – 0.546) = 0.092

Poisson distribution

The Poisson distribution may be considered as a special case of the binomial distribution where n is large and p is small, but the mean, which is equal to the product np, is a constant value. There are some situations, however, where not only is n large but it cannot be determined, e.g. errors in this book. There are many thousands of occasions when an error can occur; I can make a mistake, the printer can make a mistake, I can fail to notice the printer's mistake etc. All that we can find is the number of errors in, say, 100 pages of the book. We can then find the mean number of errors per page. As the mean is equal to np, we can thus find the product np without knowing or being able to evaluate either n or p separately.

When n is large, p is small and np is a constant value m, then it is possible to show that the formula for the number of successes given above for the binomial distribution becomes

$$P(r) = \frac{e^{-m}m^r}{r!} \text{ where } r \text{ takes the values } 0, 1, 2, 3, 4, \ldots \text{ and } e = 2.718\ 3$$

(Most books of tables include a table for e^{-x} and thus e^{-m} can be found.)
On substituting $r = 0, 1, 2, 3, 4 \ldots$ in this formula we find

$$P(0) = e^{-m} \quad P(1) = me^{-m} \quad P(2) = \frac{m^2e^{-m}}{2}$$

$$P(3) = \frac{m^3e^{-m}}{3 \times 2} \quad P(4) = \frac{m^4e^{-m}}{4 \times 3 \times 2}$$

Suppose that the mean number of errors per page is 0.4, i.e. $m = 0.4$; then on substituting in the formula we can find the probabilities of 0, 1, 2, 3, 4 ... errors per page:

Probability of no errors $= e^{-m} = e^{-0.4} = 0.670\ 3$

Probability of one error $= me^{-m} = 0.4 \times e^{-0.4} = 0.268\ 1$

Probability of two errors $= \dfrac{m^2e^{-m}}{2} = \dfrac{(0.4)^2 \times e^{-0.4}}{2} = 0.053\ 6$

Probability of three errors $= \dfrac{m^3e^{-m}}{3 \times 2} = \dfrac{(0.4)^3 \times e^{-0.4}}{3 \times 2} = 0.007\ 1$

Probability of four errors $= \dfrac{m^4e^{-m}}{4 \times 3 \times 2} = \dfrac{(0.4)^4 \times e^{-0.4}}{4 \times 3 \times 2} = 0.000\ 7$

As in the case of the binomial distribution, special tables are available which enable us to find the probabilities without so much tedious arithmetic. The tables show the probability of obtaining *r or more* successes. In the example above $m = 0.4$:

	$m = 0.4$
$r = 0$	1.000 0
1	0.329 7
2	0.061 6
3	0.007 9
4	0.000 8
5	0.000 1

Thus probability of no errors = 1.000 0 − 0.329 7 = 0.670 3
Thus probability of 1 error = 0.329 7 − 0.061 6 = 0.268 1
Thus probability of 2 errors = 0.061 6 − 0.007 9 = 0.053 6
Thus probability of 3 errors = 0.007 9 − 0.000 8 = 0.007 1
Thus probability of 4 errors = 0.000 8 − 0.000 1 = 0.000 7

We can also use the probabilities found above to find the expected number of pages in the book that contain 0, 1, 2, 3, 4 ... errors. With, say, 300 pages the expected number of pages with no errors is $300 \times 0.670\ 3 = 201$, the expected number of pages with one error is $300 \times 0.268\ 1 = 80$; similarly the expected number of pages with 2 and 3 errors are 16 and 2 respectively.

Worked Example No. 9.10

The mean number of strikes in a particular industry was found to be 1.2 per week. What is the probability that during a given week there will be (i) no strikes; (ii) more than 2 strikes; (iii) exactly 4 strikes?

$m = 1.2$

i $P(0) = e^{-m} = e^{-1.2} = 0.301\ 2$

ii $P(r > 2) = 1 - P(0) - P(1)$

$P(1) = me^{-m} = 1.2 \times e^{-1.2} = 0.361\,4$

$\therefore P(r > 2) = 1 - 0.301\,2 - 0.361\,4 = 0.337\,4$

iii $$P(4) = \frac{m^4e^{-m}}{4!} = \frac{(1.2)^4 \times e^{-1.2}}{4 \times 3 \times 2} = 0.0260$$

Worked Example No. 9.11

A manufacturer makes a component and finds that on average 4% of the components are defective. Components are packed in cartons of 40. What are the probabilities that the carton contains (i) at least one defective; (ii) exactly two defectives?

This is a problem where we could use the binomial distribution, but as n is large it is better to use the poisson distribution.

$n = 40$, p $= 0.04$, thus $m = n\text{p} = 40 \times 0.04 = 1.6$

i P(at least one) $= 1 - P(0)$

$P(0) = e^{-m} = e^{-1.6} = 0.201\,9$

$\therefore P(\geqslant 1) = 0.798\,1$

ii $$P(2) = \frac{m^2e^{-m}}{2} = \frac{(1.6)^2 \times e^{-1.6}}{2} = 0.258\,4$$

Revision Exercises

9.7 x has a standard normal distribution with mean 0 and standard deviation 1. Find the probability that x

- **a** is less than 1.2
- **b** is greater than 0.7
- **c** lies between 1.4 and 2.2
- **d** is less than −0.6
- **e** is greater than −0.4
- **f** lies between −0.5 and +0.2

9.8 x has a normal distribution with mean 4 and standard deviation 2; find the probability that x

- **a** is less than 5
- **b** is greater than 6
- **c** lies between 5.5 and 7
- **d** is less than 2.5
- **e** lies between 3.5 and 4.8

9.9 Sugar is packed in cardboard cartons and it is found that the weight is normally distributed with a mean weight of 500 g and a standard deviation of 10 g. Find the probability that a packet chosen at random has a weight.

a less than 503 g
b greater than 520 g
c that lies between 510 and 515 g
d less than 482 g

9.10 Men have an average height of 5 ft 8 in with a standard deviation $2\frac{1}{2}$ in. Assuming a normal distribution of heights, find the percentage of men

a with a height over 6 ft
b less than 5 ft 2 in
c between 5 ft 5 in and 5 ft 10 in

9.11 Jam is packed in jars with net nominal weight 500 g. The net weight is normally distributed with standard deviation 5 g. If not more than one jar in one hundred is to contain less than 500 g of jam, at what mean weight should the jars be filled to achieve this aim?

9.12 a Discuss the use and properties of the normal distribution.

b A firm makes a component with a nominal length of 10 cm. On examination the production records show that the length of the component is normally distributed with standard deviation 0.025 cm. For a component to be of use it must be between 9.96 cm and 10.05 cm. The cost of producing the component is 10p. A component of correct size can be sold for 18p. Components which are undersized are sold for scrap for 1p. Components which are oversized can be made the correct size for 3p more.

Estimate the proportion of components which are initially

i the correct size
ii oversize
iii undersize.

Hence calculate the contribution to profit (or loss) if a firm makes 1 000 components.

(H.N.D. 1976)

9.13 a A product is packed by machine. 15% of packets, however, need repacking by hand. If five packets are chosen at random from the output of the machine, calculate the probability that

i none need repacking,
ii exactly one needs repacking,
iii three or more need repacking.

b A firm which rents television sets to customers receives on average 5 calls per day from customers with a request for the television engineer to call to deal with a fault in their television set. Calculate the probability that on a given day

i no calls are received.
ii two or more calls are received.

(R.S.A. 1977)

9.14 It is estimated that in socio-economic groups AB 25% use credit cards. 10 persons are selected at random from these groups and are questioned on use of credit cards. What is the probability that (i) exactly one person, (ii) at least one person, (iii) more than 3 persons, use credit cards?

9.15 On average there are 4 calls per day made upon a company's maintenance engineers. On what proportion of days would you expect them to have no calls? Assuming there are 20 working days next month, on how many of these days would you expect them to be called upon more than 4 times?

9.16 Experience shows that 20% of customers who reserve tables at a certain restaurant do not appear. The restaurant has seats for 86 customers. The practice of the restaurant is to accept 100 reservations to allow for those that do not appear. What is the probability that the restaurant cannot seat all those who have reserved seats and appear on a particular day?

10 Confidence Intervals and Tests of Significance

In many practical situations we wish to obtain information about the population. For example the opinion polls, by asking a sample of electors how they would vote if an election were held, give an indication of how the 40 million electorate is likely to vote if a general election were held in the immediate future. In the marketing field a firm may be considering introducing a new packaging for an existing product. The firm hopes that the new packaging would improve sales. However the firm would be unlikely to change without a test marketing of the new packaging in a sample of shops, since if the new packaging does not appeal to the buying public, sales may fall and profits are likely to drop. In opinion polls and in test marketing we have little alternative but to use a sample to provide the required information.

In the case of opinion polls we are using sample data to *estimate* the proportion likely to vote for the major parties in a general election. In the case of test marketing we are using a sample of shops to examine the *hypothesis* that the new packaging will improve sales. The main use of statistical techniques is to use sample data to *estimate population values* and to test *hypotheses*.

We have a notational problem in Statistics, since we have a population value we wish to estimate e.g. the proportion voting Conservative in a general election, and a sample value, the proportion voting Conservative as given by the opinion polls. The convention we employ is to use a Greek letter to represent the population value and a Roman letter to represent the sample value. Thus the proportion voting Conservative as recorded by the opinion polls is represented by the Roman letter p, whilst the proportion voting Conservative in the electorate as a whole is represented by the Greek letter π.

p is the *sample* estimate of the *population* proportion π.

Let us suppose that we want to find the average height of male full-time students at a college. To do this we could collect every male student, measure his height, and then find the average of all these heights. The result would be the average height of the *population* of male full time students and is designated by the Greek letter μ. However, to find and measure every full-time male student would be an immense task. In practice we would choose from the registers a random sample of students (say a sample of n students) and we would measure the height of the sample and find the average of these heights. The result would be the average height of the *sample* of male full-time students and this is designated by the Roman letter $\bar{x}$.

$\bar{x}$ is the *sample* estimate of the *population* mean μ.

As $\bar{x}$ is found from a sample, it may, or may not, be equal to μ. For example, if by chance our sample contained an excess of tall men, then $\bar{x}$, the sample mean, would be greater than μ, the population mean. On the other hand, if by chance our sample contained an excess of small men, then $\bar{x}$, the sample mean, would be less than μ, the population mean. If our sample was an exact replica of the population, then $\bar{x}$ would equal μ.

When we have a sample estimate $\bar{x}$ we have to recognise that it may or may not be close to μ. The problem arises — how far or how close is $\bar{x}$ likely to be to μ? This problem is linked with the theory of probability since we need to know the probability that our sample will contain an excess of tall (or short) men. To work out these probabilities we need to know something about the distribution of $\bar{x}$, the sample mean. It is possible to show that in general $\bar{x}$ has a normal distribution. The distribution of $\bar{x}$ is called the *sampling distribution* of $\bar{x}$.

In Chapter 9 we discussed the normal distribution. In Worked Example No. 9.4 we found that for the normal distribution with mean 0 and standard deviation 1, 95% of the observations lie within 1.96 of the mean, and 99% of the observations lie within 2.58 of the mean. It is also possible to show that 99.8% of the observations lie within 3.09 of the mean. In the case of a normal distribution with mean μ and standard deviation σ we have to multiply 1.96, 2.58 and 3.09 by σ, thus 95% of the observations lie within 1.96σ of the mean μ, 99% of the observations lie within 2.58σ of the mean μ, and 99.8% of the observations lie within 3.09σ of the mean μ.

As stated earlier $\bar{x}$ has a sampling distribution which is normally distributed. We may show that the mean of this sampling distribution is μ and the standard deviation $\frac{\sigma}{\sqrt{n}}$.

N.B. $\frac{\sigma}{\sqrt{n}}$ is also called the *standard error* of the mean.

If, instead of taking one sample, we took a large number of samples each of n, we would obtain a series of values of $\bar{x}$, some larger than μ, others smaller. If the $\bar{x}$ were plotted on a graph (Figure 10.1) they would give a dotted line showing a much more 'peaked' curve than the continuous line, which is the distribution of the height of *all* students of the college.

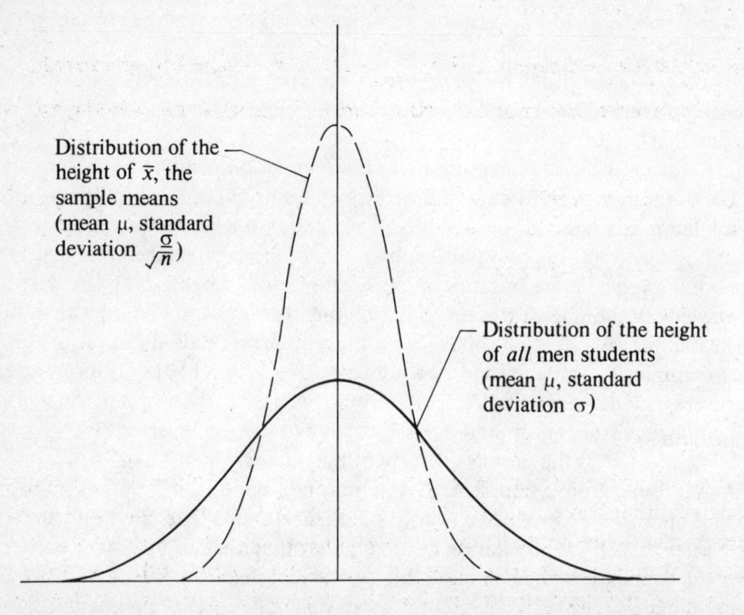

Figure 10.1

Important note

In Chapter 3 the standard deviation was denoted by the Greek letter σ. If we are given *population* data this is the correct notation, but if we are given *sample* data the standard deviation should be denoted by the Roman letter s in accordance with the convention mentioned above. Some examiners expect the correct notation to be used. If the data relate to the population, use μ for the mean and σ for the standard deviation; if the data relate to a sample, use $\bar{x}$ for the mean and s for the standard deviation.

We need to estimate the standard error of the mean $\frac{\sigma}{\sqrt{n}}$ from sample data. We obtain the sample standard deviation s, and the estimate of $\frac{\sigma}{\sqrt{n}}$ is $\frac{s}{\sqrt{n}}$.

Confidence interval for mean

Consider again the problem on heights of men students. If we want to estimate μ, the mean height of the population of men students, we use $\bar{x}$, the sample mean. As we have argued earlier, $\bar{x}$ may be an over-estimate or under-estimate of μ, according to whether we have an excess of tall men or short men in the sample. Using the theory of probability, we may say that μ is likely to be in the interval

$\bar{x} + 1.96 \frac{s}{\sqrt{n}}$ to $\bar{x} - 1.96 \frac{s}{\sqrt{n}}$ with a probability of 0.95.

$\bar{x} + 1.96 \frac{s}{\sqrt{n}}$ to $\bar{x} - 1.96 \frac{s}{\sqrt{n}}$ is called a 95% *confidence interval.*

It means that we are 95% confident that μ lies in the interval

$\bar{x} + 1.96 \frac{s}{\sqrt{n}}$ to $\bar{x} - 1.96 \frac{s}{\sqrt{n}}$

The reason for multiplying 1.96 by $\frac{s}{\sqrt{n}}$ is that if n is large (i.e. greater than 30) $\bar{x}$ (the sample mean) has a normal distribution with a mean μ and a standard deviation of $\frac{s}{\sqrt{n}}$.

We can also have 99% and 99.8% confidence intervals. We are 99% confident that μ lies in the interval

$\bar{x} + 2.58 \frac{s}{\sqrt{n}}$ to $\bar{x} - 2.58 \frac{s}{\sqrt{n}}$

We are 99.8% confident that μ lies in the interval

$\bar{x} + 3.09 \frac{s}{\sqrt{n}}$ to $\bar{x} - 3.09 \frac{s}{\sqrt{n}}$

Important note

It is very important to learn the numbers 1.96 (95%), 2.58 (99%) and 3.09 (99.8%). In some examinations, tables of the normal distribution are not provided and you have no alternative but to learn these numbers (it is therefore necessary to find out whether normal tables are supplied by the examination body). If you are given tables you may have forgotten how to find these numbers, i.e. you cannot remember how to do Worked Example No. 9.4.

Calculation of a confidence interval

To find a confidence interval we must find from the data $\bar{x}$ (the sample mean), S (the sample standard deviation) and n = numbers of individuals in the sample, and then substitute in the formulae listed above according to whether it is a 95%, 99%, or 99.8% confidence interval.

Worked Example No. 10.1

Children in households of recidivist prisoners

Number of children in the household	*Number of households*
None	25
1	40
2	42
3	28
4	18
5	11
6	9
7 or more	6
Total	179

Source: Pauline Morris, *Prisoners and their families*

i Calculate the arithmetic mean and standard deviation for the above distribution.

ii On the assumption that the above data derive from a simple random sample from a large population, calculate the 95% confidence interval for the mean number of children.

Part **i** of this question was Worked Examples Nos. 3.2 and 3.5 and we found that $\bar{x} = 2.44$ and $s = 1.926$, $n = 179$

ii 95% confidence interval is $\bar{x} + 1.96 \dfrac{s}{\sqrt{n}}$ to $\bar{x} - 1.96 \dfrac{s}{\sqrt{n}}$

i.e. $2.44 + 1.96 \times \dfrac{1.926}{\sqrt{179}}$ to $2.44 - 1.96 \times \dfrac{1.926}{\sqrt{179}}$

i.e. $2.44 + 0.28$ to $2.44 - 0.28$

i.e. 2.72 to 2.16

We are 95% certain that the mean number of children in the population of households of recidivist prisoners lies in the Interval 2.72 to 2.16.

If we had been asked to find a 99% and 99.8% confidence interval, the answers would have been

99%: $2.44 + 2.58 \times \dfrac{1.926}{\sqrt{179}}$ to $2.44 - 2.58 \times \dfrac{1.926}{\sqrt{179}}$

= 2.81 to 2.07

99.8%: $2.44 + 3.09 \times \dfrac{1.926}{\sqrt{179}}$ to $2.44 - 3.09 \times \dfrac{1.926}{\sqrt{179}}$

= 2.88 to 2.00

Exercise 10.1

Number of rooms in dwellings in Great Britain in 1975

No. of rooms	*Percentage of dwellings*
1	1
2	2
3	8
4	21
5	33
6	26
7	5
8 or more	4
Total	100

i Calculate the mean and standard deviation of the above distribution.

ii Assuming that the data is based on a simple random sample of 8 120 dwellings, calculate the 95% confidence interval for the mean number of dwellings in Great Britain.

Note. Part (i) of this question was set as Exercises 3.2 and 3.5.

Test of significance of one mean

Consider again the height of male full-time students. Suppose we want to examine whether the mean height of male full-time students is different from the mean height of all Englishmen. Let the mean height of Englishmen be denoted by μ_0. If we were to measure *every* full-time student and found the mean height μ of the population of male full-time students, we could then make a definite statement about the relative heights of male full-time students and the mean height of Englishmen.

Suppose that instead of measuring every student we chose a random sample of male students and found the sample mean $\bar{x}$. We would then compare $\bar{x}$ with μ_0. We might well find, $\bar{x}$ and μ_0 to be different. This difference could be due to *either*

a that the mean height of the population of male full-time students μ is in fact different from μ_0, or

b that the sample was not representative, i.e. by chance the sample of male full-time students has an excess of tall (or short) men.

We use the theory of probability to work out the chance that the sample was unrepresentative. As mentioned previously, we know that $\bar{x}$ is normally distributed and we can use the special tables of the normal distribution to work out the probabilities that the difference between $\bar{x}$ and μ_0 is due to **b**.

The procedure used in deciding whether the difference between $\bar{x}$ and μ_0 is due to **a** or **b** is called a *test of significance*.

We first set up the null hypothesis $\mu = \mu_0$, i.e. that the mean height of the population of male full-time students is the *same* as the mean height of Englishmen. $\bar{x}$ is of course the same estimate of μ. If the null hypothesis is true and $\bar{x}$ differs from μ_0, then this difference would be due to **b**.

We normally set up an alternative hypothesis, $\mu \neq \mu_0$, i.e. that the mean height of the population of male full-time students is *different* from the mean height of Englishmen. If the alternative hypothesis is true, then most likely $\bar{x}$ differs from μ_0 and we would agree that the difference is due to **a**.

To find out whether the null hypothesis is true we take the difference $\bar{x} - \mu_0$ and divide by the standard error of $\bar{x} = \frac{s}{\sqrt{n}}$ i.e. we calculate

$$z = \frac{\bar{x} - \mu_0}{\frac{s}{\sqrt{n}}}$$

(We take the *numerical* value of z, i.e. disregard the sign)

Now we may show that provided n is greater than 30 z is normally distributed with mean 0 and standard deviation 1.

If we find z is numerically greater than 1.96 we conclude that the difference between $\bar{x}$ and μ_0 is only 1 time in 20 due to (b); since this is a small probability, we reject the null hypothesis that $\mu = \mu_0$ and accept that the difference is due to (a), i.e. we accept the alternative hypothesis that $\mu \neq \mu_0$. We state that the difference is significant at the 5% level (1 in 20 $\doteq$ 5%) and we agree that there is some evidence of difference between $\bar{x}$ *and* μ_0, i.e. some evidence that the mean height of male full-time students is different from mean height of Englishmen.

If we find z is numerically greater than 2.58 we would conclude that the difference between $\bar{x}$ and μ_0 is only 1 time in 100 due to (b), and since the probability is very small, we conclude that the difference is due to (a). We state that the difference is significant at the 1% level (1 in 100 = 1%) and we would argue that there is good evidence that the mean height of full-time male students is different from the mean height of Englishmen.

If z is numerically less than 1.96, then there is a chance that the difference between $\bar{x}$ and μ_0 is due to (b), i.e. that the null hypothesis is true: $\mu = \mu_0$.

We may summarise the above in the following table:

Result	*Implication*	*Conclusion*	*Comment*
If z is less than 1.96	Not significant at 5% level	Accept null hypothesis	No evidence of difference between μ and μ_0
If z is greater than or equal to 1.96 but less than 2.58	Significant at 5% level	Reject null hypothesis	Some evidence of difference between μ and μ_0
If z is greater than or equal to 2.58 but less than 3.09	Significant at 1% level	Reject null hypothesis	Good evidence of difference between μ and μ_0
If z is greater than or equal to 3.09	Significant at 0.2% level	Reject null hypothesis	Very good evidence of difference between μ and μ_0

Worked Example No. 10.2

Mother's age at birth of child for a sample of mothers

Mother's age	*Number*
Under 17	1
17–19	33
20–24	282
25–29	367
30–34	250
35–39	134
40–44	41
over 44	8
Not stated	26
	1 142

i Calculate the arithmetic mean and standard deviation for the distribution.

ii The average age of all mothers (at time of child's birth) in England and Wales was 28.9 years. Test whether there is a significant difference between the average age of the sample of mothers and that of all mothers in England and Wales.

Part (i) of this question was Worked Examples No. 3.3 and 3.6 and we found that $\bar{x} = 28.98$, $s = 5.95$ and $n = 1\ 116$.
Null hypothesis $\mu = \mu_0$, i.e. $\mu = 28.9$
Alternative hypothesis $\mu \neq \mu_0$, i.e. $\mu \neq 28.9$
Our estimate of μ is 28.98, $s = 5.95$, $n = 1\ 116$
We asked if 28.98, the mean age of the sample of mothers was significantly different from the mean age of all mothers in England and Wales (28.9).

We use $z = \dfrac{\bar{x} - \mu_0}{\dfrac{s}{\sqrt{n}}}$ Here $\bar{x} = 28.98$, $\mu_0 = 28.9$, $s = 5.95$, $n = 1\ 116$

$$\therefore \quad z = \frac{28.98 - 28.9}{\dfrac{5.95}{\sqrt{1\ 116}}} = 0.45$$

$z < 1.96$; not significant at 5% level
We therefore accept the null hypothesis.

Our conclusions are that from this information there is no evidence that the mean age of mothers in the sample was significantly different from the mean age of all mothers in England and Wales.

Important note

The answer 'Not significant at 5%' by itself is not satisfactory. You *must* draw a conclusion.

Exercise 10.2

Sales (£1 000)	*Number of representatives*
Under 10	3
10 and under 20	8
20 and under 30	16
30 and under 40	13
40 and under 50	7
50 and over	3

i Calculate the mean and standard deviation of the above distribution.

ii The mean sales of representatives of another company in the same industry was £33 100. Assuming that the above data are based on a simple random sample of representatives from a company with a large number of sales representatives, examine whether there is a significant difference between the mean sales of the representatives of the two companies.

Note: Part (i) of this question was set as Exercises 3.3 and 3.6.

Confidence interval for proportion

We are sometimes interested in estimating a proportion, for example, the proportion of Conservative voters in the electorate. If we take a random sample of size n from the electoral register and of those interviewed r state that they would vote Conservative, then the sample proportion p of those who would vote Conservative is given by $p = \frac{r}{n}$

p is the *sample* estimate of the *population* value π.

When we chose our sample of electors, we could by chance have selected an unduly high proportion of Conservatives. In this case, p would overestimate π. Alternatively, by chance we could have selected an unduly low proportion of Conservatives, and in this case p would be an underestimate of π.

We may show that if n is large, then p is approximately normally distributed. We may also show that the standard error of p is $\sqrt{\frac{p(1-p)}{n}}$ a 95% confidence interval for π is

$$p + 1.96\sqrt{\frac{p(1-p)}{n}} \text{ to } p - 1.96\sqrt{\frac{p(1-p)}{n}}$$

Worked Example No. 10.3

An opinion poll was conducted by means of a simple random sample of 2 000 electors; 45% of those questioned stated they would vote Conservative if there were an immediate General Election. Calculate a

95% confidence interval for the percentage voting Conservative in the electorate.

The first step is to change the percentage to a proportion:

$$p = \frac{45}{100} = 0.45$$

A 95% confidence interval is

$$p + 1.96\sqrt{\frac{p(1-p)}{n}} \text{ to } p - 1.96\sqrt{\frac{p(1-p)}{n}}$$

$$= 0.45 + 1.96\sqrt{\frac{0.45\,(1-0.45)}{2\,000}} \text{ to } 0.45 - 1.96\sqrt{\frac{0.45\,(1-0.45)}{2\,000}}$$

$$= 0.45 + 0.022 \quad \text{to } 0.45 - 0.022$$

$$= 0.472 \quad \text{to } 0.428$$

Thus the percentage voting Conservative in the electorate is likely to be between 42.8% and 47.2% with 95% confidence.

(*N.B.* Opinion polls in practice are not based on simple random samples but on multi-stage statified samples and the standard error of the proportion is rather complicated. Firms such as N.O.P estimate that a 95% confidence interval based on a sample of 2 000 is approximately (using our figure of 45%) 45% ± 3%.)

Exercise 10.3

A random sample of 1 000 households showed 13% to be living below a defined poverty line. Calculate a 95% confidence interval for the percentage living below the defined poverty line in the population.

Significance test for proportions

Consider again our example on opinion polls. Suppose that at the last General Election the proportion voting Conservative was π_0. We now wish to investigate whether the proportion who vote Conservative has changed. If an election were held now we would of course obtain π and could make a definite statement on the degree of change. Without an election we have to rely on opinion polls and from these obtain an estimate p of the proportion voting Conservative, π.

We can argue that p may be an underestimate or an overestimate of π according to whether the sample by chance had too few or two many Conservatives.

We conduct a significance test of a proportion in a similar way to a test of a mean. We set up the null hypothesis $\pi = \pi_0$.
and alternative hypothesis $\pi \neq \pi_0$.
and calculate

$$z = \frac{p - \pi_0}{\sqrt{\dfrac{\pi_0\,(1-\pi_0)}{n}}}$$

As before, z is normally distributed with mean 0 and standard deviation 1. We compare it against 1.96, 2.58 and 3.09.

One-tailed and two-tailed tests

If market research were being conducted on a sample of mothers with young babies on behalf of a firm manufacturing goods for babies, the market research agency would hope that the sample of mothers was representative of mothers in the population. One method of checking on the validity of the sample of mothers is to test whether there is any *difference* between the mean age of mothers in the sample and the mean age of mothers in England and Wales. In Worked Example No. 10.2 the test we used showed that there was no significant *difference* and thus on the criteria of age the sample could be considered as representative of the population of mothers. When we examine for a *difference* we use both tails of the normal distribution (1.96, 2.58 and 3.09 are calculated on a two-tailed basis) and in this situation we employ a *two-tailed test.*

There are, however, some situations when it is necessary to test for a change in a particular direction. For example, if a firm wished to introduce a new packaging for one of its products it would hope that the new packaging would *increase* sales. In another situation if a firm has a problem with invoice errors it may decide to introduce a new system of invoicing and it would hope that the new system would *reduce* invoice errors. In both these examples we are interested in a change in *one* direction and in these cases we employ a *one-tailed test.*

The percentage points for one and two-tailed tests are given below:

	Two-tailed	*One-tailed*
5%	1.96	1.64
1%	2.58	2.33
0.2%	3.09	2.88

Important note

In examination problems on significance tests it is important to consider whether we need to employ a one-tailed test or a two-tailed test. If a question asks such words as 'Is there any difference?', or 'Is there a change?' then a two-tailed test should be applied. If the question uses such words as 'improvement', 'increase', 'reduction' or 'decrease', then a one-tailed test should be applied.

Worked Example No. 10.4

A firm found that 15% of invoicing had errors. The firm decided to introduce a new system of invoicing with the aim of reducing the percentage of errors. After the new system had been operating for a period a random sample of 500 invoices were checked and 61 were found to have errors. Is there any evidence that the new system of invoicing has reduced errors?

Here we have a one-tailed test, since we are examining the hypothesis that the new system of invoicing has *reduced* errors.

Null hypothesis $\pi = \pi_0$, i.e. $\pi = 0.15$

Alternative hypothesis $\pi < \pi_0$ i.e. $\pi < 0.15$ (one-tailed test)

$$\text{Estimate of } \pi \text{ is } p = \frac{61}{500} = 0.122$$

$$\text{Test is } z = \frac{p - \pi_0}{\sqrt{\dfrac{\pi_0(1-\pi_0)}{n}}} = \frac{0.122 - 0.15}{\sqrt{\dfrac{0.15(1-0.15)}{500}}}$$

$$= \frac{-0.028}{0.015\,97} = -1.75$$

We take the numerical value of $z = 1.75$.

Now as we are testing one-tailed, the 5% point is 1.64. Hence the result is significant at 5%. We therefore reject the null hypothesis and conclude that there is some evidence that the new system of invoicing has reduced errors.

Exercise 10.4

A survey based on a simple random sample showed that for a region out of 580 households questioned, 307 had a telephone. Post office records for the previous year showed that 48% of households in the region had a telephone. Is there any evidence that the percentage of households with a telephone has increased?

Significance test of two means

Suppose we want to investigate whether the mean height of polytechnic male full-time students is different from the mean height of university male full-time students.

We would take random samples from both types of institution, measure the students and find $\overline{x}_p$, s_p and n_p for polytechnic students and $\overline{x}_u$, s_u and n_u for university students.

We establish the null hypothesis $\mu_p = \mu_u$, i.e. the mean height of polytechnic and university students is the same.

We also establish the alternative hypothesis $\mu_p \neq \mu_u$, i.e. the mean height of polytechnic and university students is different.

In this situation both sample means could be affected by unrepresentative samples. For example, the null hypothesis could be true but the difference $\overline{x}_p - \overline{x}_u$ could be due to the polytechnic sample containing an excess of tall men and the university sample containing an excess of small men.

Again we use the theory of probability. This time we calculate

$$z = \frac{\overline{x}_p - \overline{x}_u}{\sqrt{\dfrac{s_p^2}{n_p} + \dfrac{s_u^2}{n_u}}}$$

Again z has a normal distribution with mean 0 and standard deviation 1 and we compare z with the values 1.96, 2.58 etc. in the case of a two-tailed test and 1.64, 2.33 etc. in the case of a one-tailed test.

Worked Example No. 10.5

The table below gives the distribution of duration of unemployment after redundancy of a simple random sample of 450 men:

Duration of unemployment	*% frequency*
Less than 1 week	46.0
1 week and under 2 weeks	9.1
2 and under 4 weeks	11.5
4 and under 8 weeks	16.7
8 and under 16 weeks	16.7
	100.0%

Source: *Department of Employment Gazette*

i Calculate the mean and standard deviation for this distribution.

ii Calculations based on a similar simple random sample of 130 women gave a mean duration of unemployment of 3.16 weeks and a standard deviation of 3.50 weeks. Examine whether there is a difference between the mean duration of male and female unemployment following redundancy.

i

Duration	*Central value* x	*Frequency* f	fx	fx^2
Less than 1 week	0.5	46.0	23.00	11.500
1 and under 2 weeks	1.5	9.1	13.65	20.475
2 and under 4 weeks	3	11.5	34.50	103.500
4 and under 8 weeks	6	16.7	100.20	601.200
8 and under 16 weeks	12	16.7	200.40	2 404.800
		100.0	371.75	3 141.475

$$\bar{x}_m = \frac{\Sigma f_x}{\Sigma f} = \frac{371.75}{100} = 3.717\,5 = 3.72 \text{ weeks}$$

$$s_m = \sqrt{\frac{\Sigma f_x^2}{\Sigma f} - \left(\frac{\Sigma f_x}{\Sigma f}\right)^2} = \sqrt{\frac{3\,141.475}{100} - \left(\frac{371.75}{100}\right)^2}$$

$$= \sqrt{31.414\,75 - 13.819\,81} = \sqrt{17.594\,94}$$

$= 4.19$ weeks

ii Null hypothesis $\mu_m = \mu_w$

Alternative hypothesis $\mu \neq \mu_w$ (two-tailed test)

$$z = \frac{\bar{x}_m - \bar{x}_w}{\sqrt{\frac{s_m^2}{n_m} + \frac{s_w^2}{n_w}}} = \frac{3.72 - 3.16}{\sqrt{\frac{(4.19)^2}{450} + \frac{(3.50)^2}{130}}}$$

$$= \frac{0.56}{\sqrt{0.039\,0 + 0.094\,2}} = \frac{0.56}{\sqrt{0.133\,2}} = \frac{0.56}{0.365} = 1.53$$

$z < 1.96$ (5% point). Thus result is not significant; accept null hypothesis. Therefore there is no evidence of difference between mean duration of male and female unemployment following redundancy.

Exercise 10.5

A new system of invoicing is designed with the aim of reducing the number of errors. To test the new system a firm selects 80 branches at random from all its branches. Half of these 80 are instructed to use the new system and the other half the old system. A careful check is made at all 80 branches to find the number of invoice errors. It is found that in branches with the new system the mean number of errors is 17.8 with standard deviation 0.5 and in the old system the mean is 18.2 with standard deviation 0.6.

Examine whether there is any evidence that the new system of invoicing has reduced invoice errors.

Significance test of two proportions

Let us consider again opinion polls. There are several opinion polls whose results are published in the national newspapers. It is not unusual for the polls to give different results even though the field work was carried out at the same time. It is quite possible for one poll to have selected by chance an unduly high proportion of Conservatives and the other poll to have selected by chance an unduly low proportion of Conservatives, and thus the poll results are likely to differ. How can we examine whether the poll results differ significantly?

We have to develop a *test of two proportions*. Suppose that in opinion poll no. 1 the sample size is n_1 and the number voting Conservative is r_1, and for poll no. 2 the corresponding figures are n_2 and r_2.

We examine the null hypothesis $\pi_1 = \pi_2$
against the alternative hypothesis $\pi_1 \neq \pi_2$

Then sample proportions are $p_1 = \dfrac{r_1}{n_1}$, $p_2 = \dfrac{r_2}{n_2}$

and to investigate whether p_1 and p_2 differ significantly we use the following test:

$$z = \frac{p_1 - p_2}{\sqrt{p(1-p)\left(\dfrac{1}{n_1} + \dfrac{1}{n_2}\right)}} \text{ where } p = \frac{n_1p_1 + n_2p_2}{n_1 + n_2}$$

Again z has a normal distribution with mean 0 and standard deviation 1 and we compare z with the values 1.96, 2.58 etc. in the case of a two-tailed test and 1.64, 2.33 etc. in the case of a one-tailed test.

Worked Example No. 10.6

Overtime working of skilled telegraphists with ten or more years of service

Overtime per year	*Day workers*	*Mixed duty and night workers*
Less than 600 hours	69	36
600 hours or more	120	125
	189	161

Source: Gwyneth de la Mare and Sylvia Shimm: *Occupational Psychology*

Calculate the percentage of workers in each of the above groups working less than 600 hours overtime per year. Test the significance of the difference between these proportions.

Describe the reasoning underlying the test and carefully interpret your results.

Percentages: Day workers $\frac{69}{189} \times 100 = 36.5\%$

Mixed duty $\frac{36}{161} \times 100 = 22.4\%$

Testing percentages is identical to testing proportions $\therefore$ change percentages to proportions (divide by 100): $p_1 = 0.365, p_2 = 0.224$

Null hypothesis $\pi_1 = \pi_2$

Alternative hypothesis $\pi_1 \neq \pi_2$ (two-tailed test)

$$z = \frac{p_1 - p_2}{\sqrt{p(1-p)\left(\frac{1}{n_1} + \frac{1}{n_2}\right)}} \text{ where } p = \frac{n_1p_1 + n_2p_2}{n_1 + n_2} = \frac{69 + 36}{189 + 161} = \frac{3}{10}$$

$$\therefore\ 1 - p = 1 - \frac{3}{10} = \frac{7}{10}$$

$$z = \frac{0.365 - 0.224}{\sqrt{(0.3)(0.7)\left(\frac{1}{189} + \frac{1}{161}\right)}} = \frac{0.141}{\sqrt{0.002\ 42}} = \frac{0.141}{0.049\ 2} = 2.87$$

$z > 2.58$

The difference is significant at 1%. We therefore reject the null hypothesis. Therefore there is strong evidence of a difference between these proportions, and we conclude that day workers and mixed duty and night workers do different proportions of overtime.

This particular question was badly worded and should have included a phrase such as 'assuming each sample to have been a simple random sample drawn from a large population ...'. In the work above we have assumed that this phrase was included in the question!

In our two samples here we have 36.5% and 22.4% undertaking less than 600 hours overtime. We want to examine whether the populations from which the samples were chosen differ. The difference between 36.5 and 22.4 could be due to two factors, (a) that the population of day workers and mixed duty and night workers do in fact have different percentages of overtime, (b) that the sample of day workers contains an excess of people doing 600 hours or less overtime and the sample of mixed duty and night workers contains too few people doing 600 hours or less overtime (both these are — or could be — due to chance).

A statistical test examines whether the difference is likely to be due to (a) or (b).

We calculate $z = \frac{p_1 - p_2}{\sqrt{p(1-p)\left(\frac{1}{n_1} + \frac{1}{n_2}\right)}}$

Now z is normally distributed with mean 0 and standard deviation 1. If z is small, i.e. less than 1.96, we conclude that the difference between sample proportions could be due to (b). If z is large, we assume that the difference between sample proportions is due to (a).

Exercise 10.6

In a random sample of 600 adults in a large town 40% are in favour of a proposal to close the centre of the town to traffic.

A group of shopkeepers opposed to the proposal carry out their own survey using a sample of 1 200 adults. Their results show 66% are opposed to the proposal.

Are these findings consistent with each other?

Revision Exercises

10.7 In the course of an audit it was found from a simple random sample of 200 bad debts that the mean debt was £48.50 with a standard deviation of £6.50.

Estimate **i** a 95% confidence interval for mean debt, **ii** a 99% confidence interval for the mean debt.

10.8 Length of residence of mortgagors in present house

Length of residence	*Percentage of mortgagors*
Under 5 years	45
5 but under 10 years	30
10 but under 15 years	11
15 but under 20 years	9
20 but under 25 years	2
25 but under 30 years	1
30 but under 35 years	1
35 but under 40 years	1

Number in sample 2 844.

Source: *General Household Survey*, 1974

a Estimate the mean and standard deviation of length of residence in the above distribution.

b In 1971 the mean length of residence of mortgagors was 7.4 years. Assuming that the data is based on a simple random sample, test whether the length of residence of mortgagors has changed.

c Explain the reasoning of the test you have employed.

(B.A.S.S. (Part-time) 1977)

10.9 In an intelligence test designed for graduates the standard score is 100 with a standard deviation of 15. A random sample of 50 polytechnic graduates were given this intelligence test and the mean score was 103.2. Is there any evidence that Polytechnic graduates differ from the standard score for graduates?

10.10 In a simple random sample of 100 farms 20% were found to have made a profit of over £3 000 in 1964–65. In 1963–64 15% of all farms made a profit of over £3 000. Is there evidence of a significant change in percentage of farms in this category?

Find a 95% confidence interval for the proportion of all farms making a profit of over £3 000 in 1964–65 and explain what the interval means.

(B.Sc.(Econ.) 1967)

10.11 Earnings of female employees, all occupations, 1965

(Women who are gainfully occupied for more than 10 hours per week)

Range of weekly earnings	*Number of employees West Midlands*
Under £6	59
£6 but under £8	30
£8 but under £10	33
£10 but under £12	22
£12 but under £14	10
£14 but under £16	2
£16 but under £18	2
£18 but under £20	3
£20 or more	7
All above ranges	168

Source: Ministry of Labour: *Family Expenditure Survey*, 1965

a Calculate mean and standard deviation for the above distribution.

b A similar distribution for the South East of 422 women had a mean of £7.90 and a standard deviation of £4.60. On the assumption that these data are based on simple random samples, test whether there is a significant difference between the earnings of women in the West Midlands Region and the South East Region.

c Explain the term 'significant' in this context.

(B.Sc. (Soc.)(Ext.) 1969)

10.12 In the course of an audit into a department of a company, a random sample of 250 records reveals 20 errors. In this department an error rate not significantly different from 4% is acceptable.

Required:

a The 95% confidence interval for the error rate in this department.

b The size of the sample which would be required to be 95% confident that the width of the estimate of the error rate for the department was not greater than 0.05.

c A test of whether the actual error rate is significantly different from an error rate of 4%.

(A.F.C. 1978)

10.13 a Explain the reasoning underlying a confidence interval.

b The Family Expenditure Survey report for 1974 shows that for Wales, of the 710 households included in the survey, 224 households had a telephone. Assuming that the data are based on a simple random sample, calculate a 95% confidence interval for the percentage of households possessing a telephone in Wales.

c In Scotland, of the 1 310 households included in the survey, 609 households had a telephone. Examine whether there is a significant difference in the proportion of households in Scotland and Wales possessing a telephone.

(H.N.D. 1976)

10.14 Household size in East Anglia in 1975

Number of household members	*Percentages*
1	18
2	32
3	20
4	19
5	7
6 or more	4

Sample Size 445

Source: *General Household Survey*, 1975

i Find the mean and standard deviation of the number of households.

ii Assuming that the data are based on a simple random sample, find a 95% confidence interval for the mean household size.

iii Explain the terms: simple random sample, sampling distribution, confidence interval.

(B.A.S.S. (Part-time) 1978)

10.15 i An investigation was carried out into the possible effect of left-handedness on typing ability. From those whose typing performance was found to be well below average, a random sample of 200 was drawn. Of these 19 were found to be left-handed.

Given that the incidence of left-handedness among the general population is 8%, do the sampling results show a significant difference in left-handedness for the poor typists?

ii A business college has a pass rate of 80% for a certain course. A large bank sends 64 employees on that course. On the assumption that they can be considered to be a random sample of all students on the course, calculate a 95% confidence interval for the number expected to pass.

Explain what you understand by the 95% confidence interval.

(H.N.D. 1977)

11 χ^2 Tests

χ^2 test of association for contingency tables

In Chapter 5 we found that the correlation coefficient is a means of examining association between two variables X and Y. The correlation coefficient, however, is suitable only when the variables X and Y can be measured on a scale, for example savings and income. Some variables cannot be measured in this way but it is often possible to set up a cross-tabulation of X against Y and to count the observation into the categories so defined.

Worked Example No. 11.1

To illustrate, consider the following example. A survey based on a simple random sample was conducted of 100 men aged 65 and over. They were asked whether they smoked and whether they suffered from bronchitis. The results of the survey are given in the cross tabulation below.

Relationship between smoking and bronchitis

	Bronchitic	*Not bronchitic*
Smokers	28	12
Non-smokers	17	43

Thus there are 28 smokers with bronchitis, 12 smokers with no bronchitis, 17 non-smokers with bronchitis and 43 non-smokers with no bronchitis.

Such a cross tabulation is called a 2×2 contingency table, since it has two rows and two columns.

We need to find out whether from this sample data in the 2×2 contingency table above there is any evidence of a relationship between smoking and bronchitis.

A quick look at the data shows that the proportion of smokers suffering from bronchitis is much higher than the proportion of non-smokers suffering from bronchitis. However as this is based on a sample there is always the risk that our sample was not representative of the population and that by chance our sample of smokers contained a large number of those suffering from bronchitis.

We need to use a test which investigates whether the difference in proportions suffering from bronchitis is due to smoking or due to the chance that our sample, though random, was not representative. We do this by setting up the null hypothesis that there is no relationship between smoking and bronchitis.

It is convenient to write the table with row and column totals.

Table 1

Observed results:

	Bronchitic	*Not bronchitic*	*Total*
Smokers	28	12	40
Non-smokers	17	43	60
Total	45	55	100

In the sample as a whole the ratio of bronchitics to non-bronchitics is 45 to 55. If the null hypothesis — that there is no relationship between bronchitis and smoking — is true, then we would expect smokers and non-smokers to be equally likely to be sufferers from bronchitis. Thus we would expect the ratio of 45 to 55 of bronchitic to non-bronchitic to apply equally to smokers and non smokers. Thus we would expect the 40 smokers to divide bronchitic to non-bronchitic in the ratio 45 to 55 and the 60 non smokers to divide bronchitic to non-bronchitic in the ratio 45 to 55.

Let us divide 40 in the ratio 45 to 55

This is equal to $\dfrac{40 \times 45}{45 + 55}$ to $\dfrac{40 \times 55}{45 + 55}$

ie. $\dfrac{40 \times 45}{100}$ to $\dfrac{40 \times 55}{100}$

ie. 18 to 22

Likewise we divide 60 in the ratio 45 to 55

This is equal to $\dfrac{60 \times 45}{45 + 55}$ to $\dfrac{60 \times 55}{45 + 55}$

ie. $\dfrac{60 \times 45}{100}$ to $\dfrac{60 \times 55}{100}$

ie. 27 to 33

Thus under the null hypothesis we would expect the following:

Table 2

Expected results:

	Bronchitic	*Not bronchitic*	*Total*
Smokers	18	22	40
Non smokers	27	33	60
Total	45	55	100

Important notes

i To find the expected results use the following method. To find the expected results for smokers who are bronchitic, multiply the corresponding row and column totals, 40 × 45, and divide by grand total, 100, viz. $\frac{40 \times 45}{100}$, and so on for each other cell.

ii As a check on your arithmetic accuracy note that the row and column totals are the same in Tables 1 and 2.

Comparing Tables 1 and 2, we note that there are differences between observed and expected results.

These differences may be due to *either*

a that there is a relationship between smoking and bronchitis; *or*

b that there is no relationship between smoking and bronchitis but the sample was, by chance, not representative of the population.

We need to use the theory of probability to decide whether the difference between observed data and expected data is likely to be due to (a) or (b).

To enable us to make this decision use the χ^2 (Chi-square) distribution. χ^2 is defined as

$$\sum \frac{(\text{Observed value} - \text{Expected value})^2}{\text{Expected value}}$$

This is written usually as $\sum \frac{(O - E)^2}{E}$

The observed values are taken from Table 1 and expected values from Table 2. For the cell smokers-bronchitic, O–E (Observed value — Expected value) is 28 − 18 = +10

We repeat this process for each cell and we find

	Bronchitic	*Not bronchitic*	*Total*
Smokers	+10	−10	0
Non-smokers	−10	+10	0
Total	0	0	0

N.B. For this table the totals are always zero.

We use the χ^2 test to examine the differences between observed and expected values. If we obtain a small value of χ^2, then the observed values and expected values are close together and we conclude that (b) is true, i.e. we accept the null hypothesis. On the other hand if we obtain a large value of χ^2, the observed and expected values are not close together and we conclude that (a) is true, i.e. we reject the null hypothesis.

Special tables, given on page 186, enable us to decide whether χ^2 is of such a value so that we may accept or reject the null hypothesis.

In this case,

$$\chi^2 = \sum \frac{(O-E)^2}{E} = \frac{(+10)^2}{18} + \frac{(-10)^2}{22} + \frac{(-10)^2}{27} + \frac{(+10)^2}{33}$$
$$= 5.56 + 4.55 + 3.70 + 3.03$$
$$= 16.84$$

We now have to work out the 'degrees of freedom'. This is an advanced concept described in books on statistical theory.

At an elementary level the idea of degrees of freedom may be considered by examining Table 2 on page 178. When we calculated that the expected result for the category 'smokers-bronchitic' was 18, the other expected results could be calculated by simple subtraction. The row and column totals in Tables 1 and 2 are the same and this imposes a constraint. Thus the expected result for the category 'smokers-non-bronchitic' is 40 – 18 = 22, the expected result for the category 'non-smokers-bronchitic' is 45 – 18 = 27, the expected result for the category 'non-smokers-non-bronchitic' is 60 – 27 = 33. It can be seen that one column and one row of expected results could be calculated. In the example on smoking and bronchitis we say we have 'one degree of freedom' since once the value of 18 was found, the other expected results followed. In the general case of an $r \times c$ contingency table (r rows and c columns) the degrees of freedom are $(r - 1)$ $(c - 1)$.

In our case we have 2 rows and 2 columns. We have therefore $(2 - 1) \times (2 - 1) = 1 \times 1 = 1$ degree of freedom.

On examining the table on page 186 for 1 degree of freedom we find

Degree of freedom	5%	2%	1%
1	3.84	5.41	6.64

Our value of $\chi^2 = 16.84$ is very much greater than the 1% point, 6.64.

There is therefore less than one chance in 100 that our value could occur due to sampling error and we conclude that observed and expected values differ significantly. We *reject* the null hypothesis (that there is *no* relationship between smoking and bronchitis). We conclude, therefore, that there is strong evidence that there *is* a relationship between smoking and bronchitis.

Larger contingency tables

The example above was based on a 2×2 contingency table. This could be made into a larger contingency table by subdividing the sample into heavy smokers, light smokers and non smokers and by also dividing men into (say) chronic, intermittent and non-bronchitic. This would yield a 3×3 contingency table. To do this would of course imply that satisfactory definitions for heavy and light smoking and for chronic and intermittent bronchitis can be set up.

Worked Example No. 11.2

Loneliness among new residents at old people's homes: by marital status

	Marital status		
New residents saying they were:	*Unmarried*	*Others*	*All*
Often lonely	18	73	91
Sometimes lonely	42	86	128
Not lonely	75	174	249
Total	135	333	468

Source: Peter Townsend: *The Last Refuge*, 1962

a On the assumption that the above data are derived from a random sample, test the hypothesis that there is no association between loneliness and marital status.

b Discuss as fully as possible the assumptions and reasoning underlying the test which you have carried out.

(B.Sc. (Econ.) 1967)

a Null hypothesis: no association between loneliness and marital status.

Expected frequencies under null hypothesis

$$= \frac{\text{Product of row + column totals}}{\text{grand total}}$$

$$\frac{135 \times 91}{468} = 26.3 \quad \frac{135 \times 128}{468} = 36.9 \quad \frac{135 \times 249}{468} = 71.8$$

$$\frac{333 \times 91}{468} = 64.7 \quad \frac{333 \times 128}{468} = 91.1 \quad \frac{333 \times 249}{468} = 177.2$$

Expected results

	Marital status		
New residents saying they were:	*Unmarried*	*Others*	*All*
Often lonely	26.3	64.7	91
Sometimes lonely	36.9	91.1	128
Not lonely	71.8	177.2	249
Total	135	333	468

Observed-Expected

	Marital status		
New residents saying they were:	*Unmarried*	*Others*	*Total*
Often lonely	−8.3	+8.3	0
Sometimes lonely	+5.1	−5.1	0
Not lonely	+3.2	−3.2	0
Total	0	0	0

$$\chi^2 = \sum \frac{(O-E)^2}{E} = \frac{(-8.3)^2}{26.3} + \frac{(8.3)^2}{64.7} + \frac{(5.1)^2}{36.9} + \frac{(-5.1)^2}{91.1} + \frac{(3.2)^2}{71.8} \frac{(-3.2)^2}{117.2} = 4.87$$

Degrees of Freedom $(r-1) \times (c-1)$ $(r = 3, c = 2)$
$(3-1)(2-1) = 2 \times 1 = 2$

From table, 2 degrees of freedom 5% 2% 1%
5.99 7.82 9.21

Our value of $\chi^2 = 4.87$ is less than 5.99, the 5% point. It is the convention to accept the null hypothesis when the value of χ^2 is less than the 5% point. (If the value of χ^2 is greater than the 5% point, we reject the null hypothesis.) We therefore accept the null hypothesis and conclude that there is no association between loneliness and marital status.

b The first assumption (which is given in the question anyway) is that data are a random sample from the population of new residents at old people's homes. The second assumption is that the interviewer did not distort the response 'often lonely and 'sometimes lonely'. Surely there are people who are on the boundary between these two conditions? In which category does the interviewer place such people? The decision in which category to place people who are not certain could materially affect the significance of the result. This is, of course, the general difficulty of 'attitude scaling'.

In carrying out a χ^2 test we make a null hypothesis, in the example above that there is no relationship between loneliness and marital status.

If the null hypothesis were true, then the residents who said they were **a** often lonely, **b** sometimes lonely, **c** not lonely, would each be divided by marital status into unmarried and others in the same ratio i.e. 135 to 333.

When we have obtained the expected frequencies under the null hypothesis and compared them with the observed frequencies, there are likely to be differences. These may be because **i** the null hypothesis is not true; **ii** the null hypothesis is true but the differences may be due to the fact that in our random sample the residents may not have been by chance representative of the population.

We calculate $\chi^2 = \sum \frac{(O-E)^2}{E}$. If 0 and E differ a lot, χ^2 is large; if O and E differ by a small amount, χ^2 is small.

We now assume that $\sum \frac{(O-E)^2}{E}$ has the χ^2 distribution so that we can use the special χ^2 tables.

These tables show how the value of χ^2 can vary by chance, and depend on the number of degrees of freedom, in this case $(r-1) \times (c-1)$.

If we have a *large* value of χ^2, say greater than the 1% point, we would argue that there is less than one chance in a 100 that the difference between observed and expected frequency is due to (ii) and at least 99 chances out of 100 that the differences are due to (i). We would therefore favour (i) and conclude that there is good evidence to reject the null hypothesis and conclude that there is a relationship between loneliness and marital status.

If we have a *small* value of χ^2, say less than the 5% point, we would argue that differences between observed and expected values are likely to be due to the sample not being representative of the population, i.e. (ii). We therefore accept the null hypothesis and conclude that there is no relationship between loneliness and marital status.

Goodness of fit tests

Important note

Some examinations include χ^2 tests of contingency tables but *exclude* 'goodness of fit' tests. It is necessary therefore to examine the syllabus and past examination papers to see whether the remainder of this chapter needs to be studied. Exercises on 'Goodness of fit' tests are found at the end of the chapter.

Goodness of fit tests involve fitting a distribution to a set of data and then testing whether the fitted distribution and the observed distribution are significantly different. At the elementary level the goodness of fit tests usually involve a Poisson distribution. Fitting a normal distribution usually takes too long to be set in an examination. The binomial distribution is not set very often; the method is very similar to that of fitting a Poisson distribution.

The method is best illustrated by means of a worked example.

Worked Example No. 11.3

The following data show the number of accidents noted by a firm during 200 working days.

Number of accidents	*Number of days*
0	112
1	64
2	18
3	4
4	2

i Calculate the mean number of accidents per day.
ii Fit a Poisson distribution.
iii Test the goodness of the fit.

i

x	f	fx
0	112	0
1	64	64
2	18	36
3	4	12
4	2	8
	200	120

$$\bar{x} = \frac{\Sigma fx}{\Sigma f} = \frac{120}{200} = 0.6$$

N.B. Examiners usually arrange that the mean is a convenient value so that tables can be used — either Poisson tables or tables of e^{-x}. If your value of $\bar{x}$ is such that you cannot use the tables, *check your calculation of the mean.*

ii $m = 0.6$ Fitted Poisson distribution has frequencies equal to $200 \times P(r)$

$$200 \times P(0) = 200 \times e^{-m} = 200 \times e^{-0.6}$$
$$= 200 \times 0.5488 = 109.8$$
$$200 \times P(1) = 200 \times me^{-m} = 200 \times 0.6 \times e^{-0.6}$$
$$= 200 \times 0.329\,3 = 65.9$$
$$200 \times P(2) = 200 \times \frac{m^2e^{-m}}{2} = 200 \times \frac{(0.6)^2 \times e^{-0.6}}{2}$$
$$= 200 \times 0.098\,8 = 19.8$$
$$200 \times P(3) = 200 \times \frac{m^3e^{-m}}{3!} = 200 \times \frac{(0.6)^3 \times e^{-0.6}}{3 \times 2}$$
$$= 200 \times 0.019\,8 = 4.0$$

At this stage note that the sum of the frequencies above is 199.5, thus the remainder is $200 - 199.5 = 0.5$

The observed and the fitted frequencies are given in the table below.

x	*Observed frequencies*	*Fitted frequencies*
0	112	109.8
1	64	65.9
2	18	19.8
3	4	4.0
4	2	0.5
	200	200.0

iii As in the case of the χ^2 test of a contingency table, we find

$$\chi^2 = \sum \frac{(O - E)^2}{E}$$

The fitted frequencies in the table above are the expected frequencies. When we use the χ^2 test the expected frequencies should not be less than 5. In the table above we have an expected frequency of 0.5, which has been joined with the group above to make the expected frequency 4.5 (this is still less than 5 but as it is very close to 5 it has been included). You should not be penalised if you use 4.5. The 'safe' way is to link the three lowest groups, the expected frequency being 24.3.

O	E	O − E
112	109.8	+2.2
64	65.9	−1.9
18	19.8	−1.8
6	4.5	+1.5
200	200.0	0

$$\chi^2 = \sum \frac{(O-E)^2}{E} = \frac{(2.2)^2}{109.8} + \frac{(-1.9)^2}{65.9} + \frac{(-1.8)^2}{19.8} + \frac{(1.5)^2}{4.5}$$
$$= 0.044\,1 + 0.054\,8 + 0.163\,6 + 0.5$$
$$= 0.762\,5$$

The null hypothesis: Poisson distribution is a good fit to the data.

Degrees of freedom — in the case of goodness of fit tests for Poisson and binomial distributions, the number of degrees of freedom equals the number of groups used in calculating χ^2 (in this example 4) less one because the total frequencies of observed and expected are the same, less another one because we calculated *m* (p in the case of the Binomial distribution) from the data. If *m* (or p) is not calculated from the data, this final degree of freedom is not lost. Thus degrees of freedom = 4 − 1 − 1 = 2

From the tables with 2 degrees of freedom χ^2 5% = 5.99. Our value of χ^2 = 0.762 5, which is less than 5.99; we therefore accept the null hypothesis and conclude that the Poisson distribution is a good fit to the data.

Worked Example No. 11.4

A large motor manufacturer is concerned about the relatively large numbers of repairs that have to be carried out under guarantee. It is decided to analyse the repair records of 200 vehicles randomly selected from the large number of jobs carried out under guarantee. From the chassis numbers it is possible to tell on what day the vehicle was made. The results were as follows:

Day of make	*Monday*	*Tuesday*	*Wednesday*	*Thursday*	*Friday*
Number of repairs	52	32	28	31	57

Is there any evidence of a significant daily variation in vehicle defects?

Null hypothesis: no difference between days in respect of vehicle defects. Under the null hypothesis we would expect equal number of defects on each day. The total number of defects is 200; thus we would expect 200 ÷ 5 = 40 on each day.

Observed	52	32	28	31	57
Expected	40	40	40	40	40
O – E	12	-8	–12	-9	17

$$\chi^2 = \sum \frac{(O-E)^2}{E} = \frac{(12)^2}{40} + \frac{(-8)^2}{40} + \frac{(-12)^2}{40} + \frac{(-9)^2}{40} + \frac{(17)^2}{40}$$
$$= 3.6 + 1.6 + 3.6 + 2.025 + 7.225$$
$$= 18.05$$

Degrees of freedom: in this case the number of degrees of freedom equals number of observations less one, since the totals of the observed and expected frequencies are the same. Thus number of degrees of freedom equals 5 – 1 = 4.

From the tables with 4 degrees of freedom χ^2 at 1% = 13.28. Our value of χ^2 = 18.05 is greater than the 1% point. We therefore reject the null hypothesis, and conclude that there is good evidence of a daily variation in vehicle defects.

Table of χ^2

Degrees of freedom	*Percentage points* 5%	2%	1%
1	3.84	5.41	6.64
2	5.99	7.82	9.21
3	7.82	9.84	11.34
4	9.49	11.67	13.28
5	11.07	13.39	15.09
6	12.59	15.03	16.82

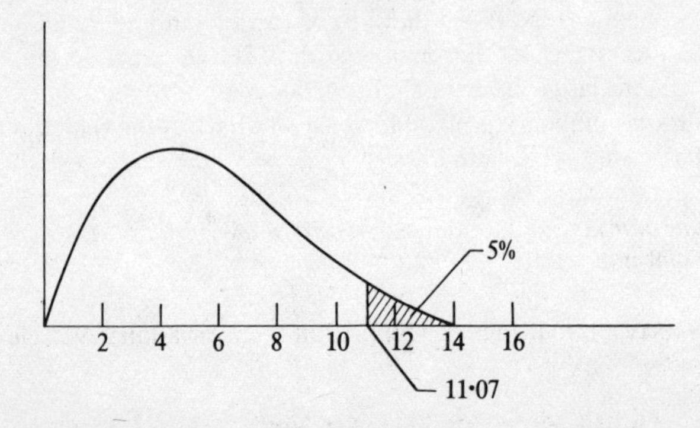

Figure 11.1

The shape of the χ^2 distribution depends on the number of degrees of freedom. In the case of 5 degrees of freedom the shape is approximately as in Figure 11.1.

The convention in using the χ^2 test is to reject the null hypothesis if the calculated value of χ^2 lies in the shaded area.

Revision Exercises

11.1

	Social Class			
	I & II	*III*	*IV & V*	*Totals*
Thumb suckers	33	49	38	120
Non-thumb suckers	99	281	220	600
Totals	132	330	258	720

Adapted from data given in J. & E. Newson: *Four years old in an Urban Community*

a Test whether the above table provides evidence to suggest that there is a significant relationship between the incidence of thumb sucking and social class.

b Explain the test you have used and the reasoning behind your conclusion in part (a).

(B.Sc.(Soc.) 1970)

11.2 In a random sample of working-class married couples with at least one parent alive, the following information was obtained:

Parents living:	*Woodford*	*Bethnal Green*	*Total*
In same borough	64	199	263
Outside borough	88	170	258
Total	152	369	521

On this evidence would you say that there are proportionately more couples living near their parents in Bethnal Green than in Woodford?

(B.Sc.(Econ.) 1961)

11.3 Briefly explain the theoretical basis for the χ^2 (Chi-squared) test.

Promotional aspirations of three occupational groups

Occupational group	*Promotion aspirations*		
	Liked the idea of promotion	*Did not like the idea of promotion*	*Total*
White-collar workers	47	7	54
Skilled manual workers	49	30	79
Semi-skilled manual workers	65	85	150
Total	161	122	283

Source: J.H. Goldthorpe et al.: 'The Affluent Worker and the Thesis of Embourgoisement', *Sociology*, Vol. 1 No. 1, Jan 1967

Do the above data give any indication that occupational category is related to promotion aspirations?

(B.Sc.(Soc.)Ext. 1967)

11.4 i Explain the reasoning underlying a χ^2 test.

ii In a firm proposals are being considered for 2 different pension plans. A sample survey to determine staff preference produced the following results:

Type of Staff	*For Plan A*	*For Plan B*	*Total*
Executive	19	23	42
Non-Executive	101	157	258
	120	180	300

Is there any indication that preference is related to job classification?

(H.N.D. 1974)

11.5 **Degree of job satisfaction**

	Very satisfied	*Fairly satisfied*	*Indifferent or dissatisfied*
Males	4 204	3 545	1 157
Females	3 430	1 757	463

Source: *General Household Survey* 1971

Required:

i Assume that the data collected is based on a simple random sample. Test the hypothesis that there is a relationship between sex and job satisfaction.

ii Briefly outline the reasoning underlying the test you have employed.

(A.F.C. 1976)

11.6 a Explain the reasoning underlying a χ^2 test.

b **Overcrowding of dwellings as measured by the 'bedroom standard'**

Occupation	*Below Standard*	*Equal to Standard*	*Above Standard*	*Total*
	Difference from bedroom standard			
Non-manual	150	1 220	3 030	4 400
Manual	510	2 410	3 680	6 600
Total	660	3 630	6 710	11 000

Source: *Social Trends* 1975

Assuming that the data are based on a simple random sample, test whether there is any relationship between occupation and overcrowding as measured by the bedroom standard. Comment on your results.

(B.A.S.S. (Part-time) 1976)

11.7 Following are **G.C.E. 'A' level results:**

Subject	*Passed*	*Failed*	*Total*
Chemistry	324	447	771
Physics	81	211	292
Applied mathematics	65	66	131
	470	724	1 194

i Is there evidence of a significant difference in pass rates between the 3 subjects?

ii Which subject, if any, is a main factor in your decision?

iii Explain carefully the bassis of this test.

(H.N.D. 1976)

11.8 **Absence from work owing to illness or accident for certain regions:**

Region	*Number absent*	*Number of employees in sample*
East Midlands	39	987
East Anglia	22	645
Wales	49	693

Source: *General Household Survey*

Assuming that the above data are based on a simple random sample, examine whether there is any relationship between absence from work owing to illness or accident and geographical region.

11.9 Television ownership by social class

Social class of head of household	*Owner*	*Non-Owner*
Upper	15	5
Middle	38	7
Lower	37	18
All	90	30

The data in the above table show television ownership by social class in a simple random sample of 120 households in Luton. Test whether there is an association between ownership of a television set and social class.

Explain carefully the basis of the test, and the meaning of your result.

(B.Sc.(Econ) 1969)

11.10 The number of vehicles passing through a particular check point during a specified time of the day were as follows:

Number of vehicles in one minute periods	0	1	2	3	4
Frequency	35	40	18	4	3

Fit a Poisson distribution to these data and test its goodness of fit.

(I.O.S. 1974)

11.11 a Describe two general applications of the χ^2 test. In each case explain how the number of degrees of freedom is determined.

b A manufacturer decides to set up an investigation to discover the effect of different-shaped bottles on sales of his product. Each type of bottle contains the same number of grammes of the product. One shape is tall and thin, the second shape is short and fat, the third is conical. A large number of bottles of each shape are placed on display. Sales are as follows:

Shape of bottle	Tall and thin	Short and fat	Conical
No. of sales	261	149	201

Examine whether there is any evidence that shape of bottle affects sales.

(R.S.A. 1973)

11.12 It is proposed to instal a set of traffic lights on a dual carriageway. A preliminary traffic count results in the data given below. Test whether these data support the highway engineer's hypothesis that the number of cars in 10 second intervals follows a Poisson distribution with mean 0.5.

Cars passing in 10 seconds	*No. of occasions observed*
0	63
1	24
2	10
3	3

12 Survey Methods

In Chapters 1 to 11 we have seen the various methods of presenting and analysing data. How were the data obtained? In most cases Government publications were the source of the data — data collected by someone else is called *secondary data*. If we want to obtain data which have not been published, we have to collect the data ourselves (*primary data*). Data collection is a big subject and in the space of one chapter it is not possible to be comprehensive. The main areas are outlined and in the appendix model answers to a number of standard questions are given. For a comprehensive coverage of this subject the standard British work is *Survey Methods in Social Investigation* by C. A. Moser and G. Kalton, published by Heinemann.

A sample survey involves the following steps:

i Stating the objectives of the survey.
ii Obtaining finance and resources to conduct the survey.
iii Collection of the data:
- **a** Compile or obtain sampling frame;
- **b** Decide on sampling method;
- **c** Decide on method of collection — interviewing, postal questionnaire or observation.

iv Definition of units.
v Design of questionnaire.
vi How to deal with non-response.
vii How to check errors.
viii Whether to have a pilot survey to examine the effectiveness of steps (i) to (vii).
ix Organise the analysis and interpretation of the collected data.

Objectives of the survey

It is clearly necessary to decide precisely what are the objectives of the survey. Sometimes the results of the pilot survey mean that the objectives need to be reassessed.

Resource implications

Data collection costs money. A small survey by students in a college on such topics as library facilities or canteen arrangements is relatively inexpensive — the costs of paper and printing are inevitable, postage costs are usually avoided and the time spent on distribution of questionnaires, collection and analysis is usually done by students at effectively no cost. However, in a survey which involves employment of interviewers, printing, postage, checking interviewers' recording schedules, follow-up of people who were out when the interviewer called, analysing results, computer time etc. the costs mount very rapidly. With limited finance and with limited voluntary help the survey has to be tailored to the resources available.

The collection of data

Suppose we wish to obtain information about a population (in statistics 'population' refers to the body from which we want to obtain information, the population could be all students at a college; all employees of a company; all people in a town; all public companies etc.) We could try to obtain information from all members of the population. This is done in the decennial population census, but this is a very big and expensive operation and some of the results take years to be published. In most cases the collection of data has to be based on a sample. A sample survey is very much cheaper and usually the results can be published quickly. If data are obtained from a sample, it is essential that the sample should be representative of the population. A sample may, for various reasons, be unrepresentative of the population; in such circumstances data collected from the sample may give a misleading impression of the population. Great care, therefore, must be used to choose a suitable sample.

Methods of choosing a sample are in two main categories:

a Random sampling
b Non-random sampling

In Chapter 10, when we discussed hypothesis testing and confidence intervals the reasoning was on the basis of random sampling; thus to use the techniques of Chapter 10 we have to have random samples. If we have a non-random sample, it is not possible to obtain a confidence interval for, say, the population mean. The criterion of randomness in sampling is that every individual in the population has an ascertainable probability of being selected. When we discuss the various methods of sampling it will be seen that the probability of an individual being selected depends on the sampling method.

Sampling frame

Random sampling requires that a list of the population is available so that the sample can be chosen at random from the population. Such a list is called a *sampling frame*. A sampling frame could be the names of all full-time students of the polytechnic, or the names of all companies which have a stock exchange listing.

Types of sampling frames

1 Lists of individuals used for administrative purposes.
2 Lists of dwellings within areas, or of other units, e.g. shops, factories.
3 Census returns from a completed census.
4 Returns from a sample survey.

In Great Britain there are two national sampling frames available for the human population, (1) a list of dwellings — the rating records, (2) a list of individuals — the electoral registers.

The electoral register

The electoral register is compiled each October, is published in February and is used for one year. It is four months out of date on publication and 16 months out of date just before the new register is published. Voters who become 18 before the register ends are marked with date at which they become êligible to vote. Evidence suggests that about a quarter of such voters are not registered and that about 96% of eligible electors are living at the qualifying address on compilation day, about 93% on publication and 85% at the end of the life of the register.

The electoral register is centrally available. The British Museum has a complete set. The electoral register for a particular constituency is usually available at a central library in that constituency.

The electoral register is extensively used as a sampling frame by government, market research and academic researchers.

The rating record

Local authorities maintain records of properties in their areas. There is no central register of rating records. There is no uniform method of keeping the records and permission to use the record for sampling purposes is at the discretion of the Rating Officer. Authorities vary in the speed at which they amend their records.

The rating records can be used as a sampling frame for dwellings, but non-dwelling property (commercial and industrial) has to be excluded.

Ideally, a sampling frame should have the following properties:

1. *Adequacy*. Does it cover the whole population to be surveyed? The electoral register excludes all persons aged under 16 years 8 months.
2. *Completeness*. Does it include every unit of the population which it claims to list? Is every one entitled to vote on the electoral register?
3. *Accuracy*. Is the information correct? Does it include people who have died or have moved?
4. *No duplication*. Is there any duplication? Students often appear on twc electoral registers and thus have greater chance of selection.
5. *Topicality*. Is the list up to date? Do the rating records include houses which have recently been completed?
6. *Convenience*. Is the frame readily accessible? Can it be used for stratification and cluster sampling?

The simple random sample

In this type of sample, the sample units are drawn directly from the whole population by some procedure which is designed to meet the essential criterion of randomness mentioned above.

The method of selecting a simple random sample involves numbering every individual in the population. Then, by randomly selecting from cards or counters which are similarly numbered, or by using random number tables, the requisite number of individuals from the population are obtained.

Simple random samples have a number of drawbacks.

1. The selected items are subject to the full range of variation of the population.
2. A very unrepresentative sample may result. For instance, a simple random sample of towns could turn out to consist entirely of new towns.

3 It may yield a selection of individuals with a wide geographical scatter, thus adding to the expense and difficulty of getting the information.

4 The numbering of the population may prove very laborious. Furthermore there might not be an adequate sampling frame.

From Chapter 10 we found the confidence interval for the population mean μ based on a simple random sample of size n: $\bar{x} + 1.96\frac{s}{\sqrt{n}}$ to $\bar{x} - 1.96\frac{s}{\sqrt{n}}$ The precision of a sampling method depends on how close $\bar{x}$ is to μ. In a simple random sample, since s is fixed by the nature of the data, precision can be improved only by increasing n the size of the sample. It should be noted that as the standard error equals $\frac{s}{\sqrt{n}}$ by increasing the sample size by a factor of four i.e. the sample size n is increased from n to $4n$, the precision is improved by a factor of two, since

$$\frac{s}{\sqrt{4n}} = \frac{s}{2\sqrt{n}}$$

The advantages and disadvantages of other forms of random sampling methods are judged by comparison with the properties of simple random sampling.

Systematic sample (interval sample)

This method is based on systematic sampling from lists. If a sample of 50 is wanted from a list of 5 000, then a number between one and 100 is randomly chosen, say 24, and 24th, 124th, 224th etc. items are selected.

If the lists are in random order (eg alphabetical list of students), a systematic sample is essentially the same as a simple random sample. However, a systematic sample does not fully meet the criterion of randomness since some samples of the given size have zero probability of being chosen. However, because of its ease and cheapness this method is widely used.

Stratified random sample

Some of the drawbacks of simple random sampling can be avoided if, as a preliminary to selecting the sample, the population is 'stratified'. This simply means that the population is divided into suitable sub-populations, called strata.

Simple random samples or systematic samples are then chosen from each stratum in such a way as to obtain a total sample of the required size. Strata are nearly always multiple classifications. In social surveys, for instance, there is usually stratification by age, sex and social class. This implies that the sampling frame must contain information on these three variables before this threefold stratification of the population can be made. Stratification variables or factors must have relevance to the purposes of the survey. In a survey of deviancy, there would be no point in stratifying by hair colour.

Stratification ensures a representative sample, since it guarantees that every important category will have elements in the final sample. If the same proportion of individuals is chosen from each stratum, then the proportionate structure of the sample will exactly reflect that of the population. Moreover, since each stratum is represented by a randomly chosen sample, it can be treated as a domain of study in its own right and valid inferences can be made about that stratum in the population.

The main statistical purpose of stratification is to increase precision. This depends, inter alia, on the variability in the population being sampled. With stratification, sampling takes place *within* strata, where the range of variation is less. Variation *between* strata does not enter as a chance effect. Hence the gain in precision.

The formula for the standard error of a stratified random sample is complicated. In a simple random sample the standard error is $\frac{s}{\sqrt{n}}$. In essence, as stratification reduces variability, s is smaller than in the case of simple random sampling and therefore there is a gain in precision.

It follows from this that the gain from stratification is greatest if individuals within a stratum are as alike as possible while the differences between the strata are as great as possible. If stratification fails to partition the population in this way, then the stratified sample is equivalent to — but no worse than — a simple random sample.

Stratification obviously requires prior knowledge of each individual in the population. Sampling frames do not always carry this information. For example, it is impossible to classify the entries in the U.K. electoral register by age because individual voters' ages are not given.

Cluster and multi-stage random sampling

Populations often exist as a hierarchy of groups or 'clusters'. In the U.K. there are parliamentary constituencies, each of which is a 'cluster' of wards. Each ward is a 'cluster' of addresses, and each address is, in turn, a 'cluster' of individual voters.

To obtain a sample of voters, we could first take a random sample of constituencies, then a random sample of wards from each sampled constituency, then a random sample of addresses would make up the final sample. This would be an example of a multi-stage sample design using clusters as sampling units at each stage except the last.

The advantages of such a procedure are clear. The field work is concentrated on the constituencies selected at the first stage. The laborious numbering of every elector in the country and the wide geographical scatter which would result from a simple random sample are avoided. Moreover, multi-staging and stratification can be and usually are combined. The clusters are stratified at each stage before the random selection is made.

Unfortunately, cluster sampling usually has an adverse effect on precision. A cluster is often a natural grouping of similar individuals. Thus, at some stage, there is often a marked difference *between* clusters. Thus one ward may be mostly fashionable, expensive addresses; another ward may be predominantly working class. Even if wards are stratified, some of this extreme variation is nearly always reflected in the sampling. Ideally, clusters should be as mixed as possible *within* and there should be as little difference as possible *between* clusters: the population should be well mixed.

A multi-stage sample usually has a larger standard error than a simple random sample of the same size. In a multi-stage stratified sample there is a gain in precision because of stratification and a loss in precision because of a multi-stage design; the *net* effect is usually a loss of precision compared with a simple random sample of the same size. The loss of precision does depend on the clustering but a standard error of a multi-stage stratified sample of about 1.5 times that of a simple random sample of the same size is fairly typical. In economic terms precision has to be considered in relation to the relative cost of multi-stage stratified and simple random sampling. If a survey requires interviewers a multi-stage stratified sample of 2 000 voters in the U.K. is very much cheaper than a simple random sample of 2 000 voters in the U.K. Thus for the same budget a simple random sample would have to have a sample size very much less than 2 000. We recall that for a simple random sample the standard error is $\frac{s}{\sqrt{n}}$ and a smaller sample means n is smaller and thus there is a reduction in precision. Hence for the *same* expenditure a multi-stage stratified sample is likely to be more precise.

Quota sampling

In this method of drawing samples, the essential randomness is forfeited in the interests of cheapness and administrative simplicity. Quota samples may incorporate stratification and multi-staging in the sample design but the selection of the final stage units is left to the interviewer. In probability sampling, the final stage units, by some randomising procedure, are selected in advance and the field-worker has no discretion.

Quota samples are, however, by no means haphazard. They are designed to ensure that the total sample mirrors the structure or stratification of the target population. Depending on this structure, the interviewer is told to interview people in certain categories. For example, if an interviewer were asked to interview 20 people then the twenty people might have to fall into the categories below:

Age	*no.*	*Sex*	*no.*	*Social class*	*no.*
18–29	5	Male	10	AB	3
30–44	6	Female	10	C1	6
45–64	7			C2	6
65 +	3			DE	5

The numbers in each category reflect the situation in the town or constituency; for example if the town were a seaside resort the number to interview in the 65 + group would be higher than 3 since seaside resorts attract retired people, likewise a constituency in a very poor area would have few people in class AB and more in class DE and numbers to be interviewed in these classes would have to reflect this situation. The numbers in each category are called the quota controls. (N.B. The social class categories are usually defined as follows: AB = managerial, professional, executive; C1 = clerical, C2 = skilled manual, DE = semi-skilled and unskilled manual.)

Some surveys allow the interviewer to select freely within these categories; other surveys give the interviewers much more specific instructions i.e. interview 20 people in the following categories (1) Male, C1 aged 30–44 (2) Female, AB, aged 65 + (3) Female, DE, aged 18–29 and so on for 20 people.

As already mentioned, the chief advantage of quota samples is their relative cheapness and administrative ease. A much larger sample and more information can be obtained, more quickly, for a given outlay than with a fully randomised sample. No sampling frame or enumeration of the population is needed. In some cases, such as television audience research, only quota sampling is feasible. Given suitable, trained and properly briefed fieldworkers, quota samples yield enough accurate information for many forms of commercial market research.

It is true that the method is known to result in certain biasses; however, these can often be allowed for, and may in any case be unimportant for the purpose of the research. It is also true that the non-random nature of the method rules out any valid estimate of the sampling error in estimates derived from the sample. Against this it can be argued that when other large sources of error, such as non-response, exist it is futile to worry overmuch about sampling error. Nevertheless, quota sampling cannot be regarded as ultimately satisfactory in research where it is important that theoretically valid results should be obtained.

Random walk

Quota sampling is used mainly for interviewing in the street. For some surveys contact with people in houses is necessary, i.e. to test a product such as a washing powder, where it is necessary to call twice, first with the washing powder and secondly to find the results. In random walk the interviewer is given a starting point — usually a road junction. The interviewer is told in which direction to proceed and then follows a route specified by instructions such as take the first turning on the left, then the first turning on the right, then the first turning on the left and so on. Along this route the interviewer is asked to call at (say) every 5th house. For random walk sampling no sampling frame is necessary but as the sample is not truly random no estimate of standard error is possible.

Methods of collecting data

The remarks made above assumed that information was to be obtained by means of personal interviews. There are two other methods of collecting data; one is to send questionnaires by post, the other is to obtain information by means of observation.

Observation

Observation has been used by some sociologists, particularly in areas where interviewing and the postal questionnaire are not suitable e.g. studying behaviour. Observation has been used to study effects of safety campaigns on pedestrians and other road users. Another form of observation is participant observation where the observer joins the group, company or organisation he is studying. He can see how the community behaves and reacts to events as they occur. The participant observer needs considerable skill. There is, however, a risk that he may obtain a restricted view. He obtains more information from the areas, the group or organisation into which he comes into contact, but the people or areas he has less (or no) contact with might have given him a completely different view. It must be recognised that his contacts may not be representative and the results may suffer accordingly.

Postal questionnaire

An alternative method is to send the questionnaire to the respondents through the post. An immediate advantage of postal questionnaires is their cheapness. An immediate drawback is the much lower response rates generally achieved.

There are other subsidiary gains from a postal survey. For a very widely scattered target population, this may be the only feasible method. The degree of clustering and the resultant loss of precision need not be so high. Interviewer bias is avoided. Considered rather than off-the-cuff responses can be hoped for. Information may be elicited which it would be embarrassing to obtain in a face-to-face interview with a stranger.

There are likewise subsidiary limitations. It is essential that a mailed questionnaire be straightforward and readily understood by those to whom it is directed. The replies have to be accepted as they come in. No subsequent elucidation is possible. There can be no guarantee that the schedule has been completed by the addressee without assistance from or discussion with others. The respondent sees the whole questionnaire before he starts filling it in, so that his responses are not independent.

Definition of units

It is essential to have a clear idea of the units to be used in the survey, for example is the unit a person, a family or a household? Whichever unit is chosen, it must be clearly defined. How is a person who lives in a house with other people but is a lodger or lives in a bed-sitter to be considered? Does such a person form a separate household?

Questionnaire design

Whether the method of collection is a postal questionnaire or interviewing, a list of questions must be formulated. The list of questions sent through the post is called a *questionnaire*; the list of questions asked and recorded by the interviewer is called a *recording schedule*. It must be remembered that the questionnaire will be completed by the responder unaided, whereas the recording schedule will be used by an expert interviewer. The design of a questionnaire or recording schedule needs to take this difference into account. The following notes are in most cases equally applicable to the questionnaire and the recording schedule.

Some general points

1. The objects of the survey should be defined as precisely as possible and the questions should aim to meet these objects.
2. Questionnaires should not be too long.
3. Factual questions about age, occupation, etc. should come at the beginning or the end.
4. The sequence of questions should have some kind of natural logical sequence which the respondent can appreciate. This will help to sustain interest. An example of this is the so-called 'funnel' approach in which the opening

questions are of a very general nature, with succeeding questions becoming more and more specific.

5 For some respondents, the questionnaire may be shortened by the use of 'filter' questions, e.g. if the answer to the 'filter' question is negative, then some of the following questions can be skipped.

Question content

Questions may be thought of as being *open* or *closed*, *opinion* or *factual*.

Open: respondent expresses his own views in his own words.

Closed: the respondent is presented with a checklist of possible answers. Alternatively both these features may be combined in the same question.

In an interview, the answers to open questions have to be recorded by the field-worker verbatim or as fully as possible in the time and space available. This is not always easy. Furthermore, the opinions so recorded are difficult to quantify, classify and code. On the other hand, open questions yield a richness and spontaneity and help to sustain the interest of the respondent. The interviewer may also have opportunities for 'probing'.

With a checklist of closed questions, there is always the possibility that the list may exclude something of substance which the respondent would wish to communicate. Equally, it might suggest things which would never have occurred to the respondent. There is also the risk that the mere underlining or ticking on a list can become too mechanical and may lack spontaneity of response. As against these drawbacks, closed questions are easy to count, classify and code.

Clearly, open questions are more appropriate for sounding out opinions, whereas closed questions are better adapted to ascertaining facts.

In framing 'opinion' questions, the following points are worth mentioning:

1 Has the respondent ever formed an opinion on the topic asked about. Has he any knowledge of it?
2 Is it necessary, desirable or possible to measure the 'intensity' with which he holds an opinion?
3 How sensitive is the expression of an opinion to slight changes in wording and emphasis in the question?
4 With regard to 'factual' questions, is the respondent likely to have knowledge of the factual matter asked about without reference to records or to other persons?

Question wording

1 Questions should not be worded too generally, but should be as specific as possible. Instead of asking 'What do you think about examinations?' questions should be asked as to whether examinations are a stimulus to study, whether they really measure attainment, and so on.
2 Keep the language of the questions as simple as possible. Avoid jargon, technical language, learned words. Make sure that words mean the same thing to the respondent as they do to the researcher, e.g. in some parts of the U.K. the words 'book' and 'magazine' are interchangeable.

3 Avoid leading questions of the type 'Have you stopped beating your wife?'
4 Where a time-scale is involved, avoid vague words such as 'generally' or 'sometimes'. Ask instead 'How many times in the last week, month, etc.'
5 Avoid biased questions such as 'Would you agree that trade unions are too powerful?'
6 Avoid 'presuming' questions: establish that the respondent is a Christian before asking about his churchgoing.
7 Avoid hypothetical questions such as 'Would you be willing to pay for private medicine if the National Insurance contributions were reduced?

Non-response

In surveying human populations, whether by sample or by attempting a complete census, there is always some degree of non-response.

If the non-responders were similar in all respects to the responders, there would be no problem apart from the reduction in the sample size. Experience shows, however, that the missing part does often differ materially from the rest, and certainly it is never safe to assume that this will not be the case. No one can say with certainty what response rate will invalidate inferences about the population. For this reason alone, the survey design must incorporate procedures for minimising non-response and then for obtaining as much information as possible about the irreducible residue of non-respondents.

In randomised surveys, Moser and Kalton distinguish five main sources of non-response and suggest ways of dealing with them.

1 *Unsuitable for interview* — infirm, deaf, foreign, etc. Little can be done in these cases.
2 *Movers*. Some of the selected individuals will have moved by the time the fieldwork is done. Subject to certain safeguards, the new household may be substituted. It may be possible to trace the person originally selected, at his new address, without too much trouble.
3 *Refusals*. The surest way of minimising refusals is to employ only carefully-selected, well-trained professional interviewers who can make a favourable impression when the crucial first contact is made on the doorstep. Reputable sponsorship of the research, and a well-designed schedule of questions which is not too long will also help to keep refusals down.
4 *People not at home* at time of call. This can be kept down by not calling at times when the respondents are likely to be out, e.g. during working hours. At least one call-back should be made, possibly on a sub-sample of the not-at-homes.
5 *People who are away from home* during the period of the fieldwork. In this situation, call-back is often impracticable and these non-responders must be written off as lost.

Nevertheless, even if every effort is made at the time, when the fieldwork has been completed there will still be an irreducible residue of non response, and the non-responders may well differ from the responders in some way which has a bearing on the research. Bias will inevitably result. This bias can only be corrected by gaining some knowledge about the non-respondents. Often the interviewer can find out from other members of the household some of the characteristics of the selected informant — age, occupation, family size — and can make a fair guess at social class, and fieldworkers should be instructed to do this kind of thing as a matter of routine. The composition of the achieved sample should be compared with known published information about the population as a possible way of identifying biasses. It may be possible to get some of the non-responders to answer a few simple questions, even if they were not prepared to submit to the full interview questionnaire. Such devices, whilst they can give some reassurance, cannot of course answer the crucial question — are the non-responders different in respect of the objects of the survey?

If an achieved sample is found to be unrepresentative in some respects, it is usual to adjust the results. The aim of a sample is to estimate certain population characteristics as accurately as possible, and it would be irrational to leave a sample wrongly constituted when one has the knowledge and resources to improve it. The adjustment is usually done by 're-weighting'. Thus, if the sex proportions in the achieved sample turned out to differ from the known sex proportion in the population, then in calculations involving this proportion, population sex data would be used instead of the sample data. This at least ensures that the sexes are correctly represented in the sample. It will not of course remove any sex-related bias in the sample data themselves.

Non-response in mail questionnaires

One reason for low response rates in postal surveys is that the target population is inherently unsuitable for approach through the post. Careful thought should be given to this before the postal method is decided upon. Empirical evidence suggests that the sponsorship of the survey is of some importance.

How to check errors

There are three main types of errors which occur in surveys.

1 *Recording errors*, e.g. when the respondent gives the answer 'no' the interviewer accidentally enters 'yes'. Errors of this type are inevitable but are not usually serious because they tend to cancel out.
2 *Interviewer bias*. On some occasions an interviewer either consciously or unconsciously affects the respondent's answers. Such bias tends to be in one direction and thus is a much more serious error than (1). Extensive training of interviewers reduces such errors. Re-interviewing a sub-sample of respondents by a supervisor helps to detect the presence and degree of interviewer bias (re-interviewing also helps to monitor recording errors).
3 *Untruthful answers*. In the case of the Family Expenditure Survey, it is known that consumption of alcoholic drink and tobacco is understated. A check with Customs and Excise statistics on the amount of alcohol and tobacco taken out of bond shows a serious shortfall. The Family Expenditure Survey is used to find the weights for the General Index of Retail Prices; therefore in finding the weights for alcoholic drink and tobacco the Family Expenditure Survey results have to be adjusted to deal with the known shortfall. There are also problems with questions on sensitive topics such as sex, or anti-social behaviour. In the case of the number of baths per week that people say they have, most reservoirs would be empty if the answers were true! To try to detect untruthful errors in a

particular questionnaire is difficult but a check on the survey results with other known results helps to identify areas where the answers are likely to be incorrect.

The pilot survey

It is nearly always better to run a small exploratory or pilot survey before plunging into a large-scale exercise. Such a pilot survey could be useful in the following ways:

1 It could reveal deficiencies in the sampling frame and in the sample design generally.

2 It could provide guidance as to whether the postal or the interview method is better.

3 It could give an indication as to the likely response rate in a larger survey. It might also help to identify the main sources of non-response and indicate how they might be dealt with in the main survey.

4 It will test out the first draft of the questionnaire. The answers to open questions in the pilot could provide the basis for a checklist of closed questions in the main survey.

5 It could provide insights into the problems facing the interviewers.

6 It could provide guidance as to the best form of administrative organisation and documentation for the survey.

Analysis and interpretation of the data

The first step is to extract the data from the questionnaire. The questionnaire should be designed to allow for speedy transfer of data to punched cards; provision should also be made to detect errors in the questionnaires. When we have extracted the data we are in a position to use chapters 1 to 11, which explain how to analyse and interpret the data.

Revision Exercises

12.1 You have written a report on a plan to establish a market research department in a company. The managing director of this company is not a statistician, has read your report and has made notes of some terms which are not familiar to him:

a Quota sampling and simple random sampling.
b Non-response.
c Sampling frame.
d Multi-stage stratified sampling.

Write reports on *three* of these terms so that the meaning and use will be clear to the managing director.

(H.N.D. 1976)

12.2 Write notes on two of the following:

a The relative advantages and disadvantages of the personal interview and the postal questionnaire as a means of collecting data.

b The basic principles of questionnaire design.

c The requirements for a good sampling frame, illustrating your answer with reference to the electoral register.

d The relative merits of the various methods of sampling.

(B.A.S.S. (Part-time) 1977)

12.3 'Simple random sampling is the best method of sampling'. Discuss this statement.

(A.F.C. 1976)

12.4 Describe and compare the following main types of sample design, quoting the type of survey for which each might be appropriate:

Simple random;
Stratified;
Multi-stage/cluster;
Quota.

(H.N.D. 1975)

12.5 'Pilot' surveys are standard practice in professional survey enquiries. Explain their purpose, giving details of the way in which such a pilot enquiry could help in carrying out the main survey.

(H.N.D. 1975)

12.6 You are asked to conduct a survey of a town of about 50 000 inhabitants to find out the extent of the ownership by households of consumer durable goods (washing machines, refrigerators, tumbler dryers, irons, vacuum cleaners, etc.). You have a limited budget for this project. Set out in some detail how you would conduct the survey.

(R.S.A. 1975)

12.7 A new town corporation wishes to conduct a survey of the town to find the views of the inhabitants on the need for a leisure centre and the facilities to provide in such a centre.

The corporation decide to question about 600 inhabitants using interviewers calling at dwellings. The council wishes to conduct the survey at minimum cost.

The new town has a population of about 60 000 and the accommodation is as follows:

Owned by new town corporation
1 000 ground-floor flats for the elderly
1 000 semi-detached houses
4 000 terraced houses
2 000 flats in low-rise buildings (6 floors or less)
4 000 flats in high-rise buildings (7 floors or more)
(The above consists of 6 estates and each estate has a mixture of dwellings.)

Privately owned
2 000 town houses
4 000 semi-detached houses
1 000 detached houses (3 bedrooms)
1 000 detached houses (4 bedrooms or more)
(The above consists of 10 estates and each estate has a mixture of dwellings.)

a Explain how you would find a sample using two of the following methods:

i Stratified sampling;
ii Multi-stage sampling;
iii Stratified multi-stage sampling.

b Which methods would you recommend the corporation to use? Give reasons for your choice.

(B.A.B.S. 1977)

Solutions to exercises

Chapter 1

1.1

Seconds	*Number*
40 - 49	25
50 - 59	32
60 - 69	18
70 - 79	11
80 - 89	8
90 - 99	6
Total	100

1.3 and 1.4

Note: as the times are given to the nearest second the class intervals are $39\frac{1}{2} - 49\frac{1}{2}$, etc.

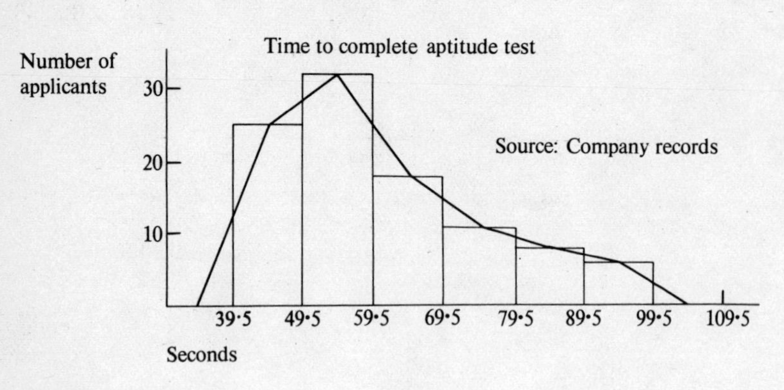

Figure A1.3

1.2 a Tenure of dwellings in England and Scotland in 1971 (thousands)

	England		*Scotland*	
	No.	%	No.	%
Owner-occupied	8 360	52.02	540	30.00
Rented —				
local authorities	4 500	28.00	940	52.22

	England		*Scotland*	
	No.	%	No.	%
Rented — private owners	2 410	15.00	200	11.11
Other tenures	800	4.98	120	6.67
Total	16 070	100.00	1 800	100.00

Source: *Social Trends* 1972

b There is a marked difference in the percentage of owner-occupiers in England and Scotland. In England owner-occupiers exceed 50% of the total; in contrast in Scotland dwellings rented from local authorities exceed 50%. These data show that there is a marked difference in the tenure of dwellings in the two countries.

1.5 The percentages to use in a percentage bar chart are given in the table for the solution of Exercise 1.2.

The angles for the pie chart are given below.

	England	*Scotland*
Owner-occupied	187.3	108.0
Rented — local authorities	100.8	188.0
Rented — private owners	54.0	40.0
Other tenures	17.9	24.0
Total	360.0	360.0

1.6 *Ordinary graph paper*

Advantages

i Shows the actual change in the variable.

ii Zero and negative values can be shown.

iii Ordinary graph paper is well known and it is easy to use.

iv Can be used to show the constituent parts of a total.

v An upward sloping straight line shows an increase by constant amounts, i.e. a simple interest rate of growth.

Disadvantages

i When there are large differences in the data, e.g. if in January 1968 there were 10 000 colour TVs licences, and in May 1978 there were 11 148 000, this would be difficult to show on ordinary graph paper.

ii Rates of change with time can be compared only by looking at gradients.

Logarithmic graph paper

Advantages

i Shows the rate of change in the variable.

ii Very useful when there are big differences in the variables.
iii An upward sloping straight line graph shows an increase at a constant rate, i.e. a compound interest rate of growth.

Disadvantages

i Zero and negative values cannot be shown.
ii More difficult to use and if logarithmic graph paper is not available, the logarithms of the variables have to be found.

1.7.

Cumulative no. of tax units	*Cumulative income*	*Percentage cumulative numbers*	*Percentage cumulative income*
3 579	1 800	12.7	3.5
9 078	6 313	32.3	12.1
13 566	11 874	48.2	22.7
17 786	19 234	63.2	36.8
21 586	27 743	76.8	53.1
24 350	35 277	86.6	67.6
27 464	46 588	97.7	89.2
27 987	50 036	99.5	95.8
28 123	52 219	100.0	100.0

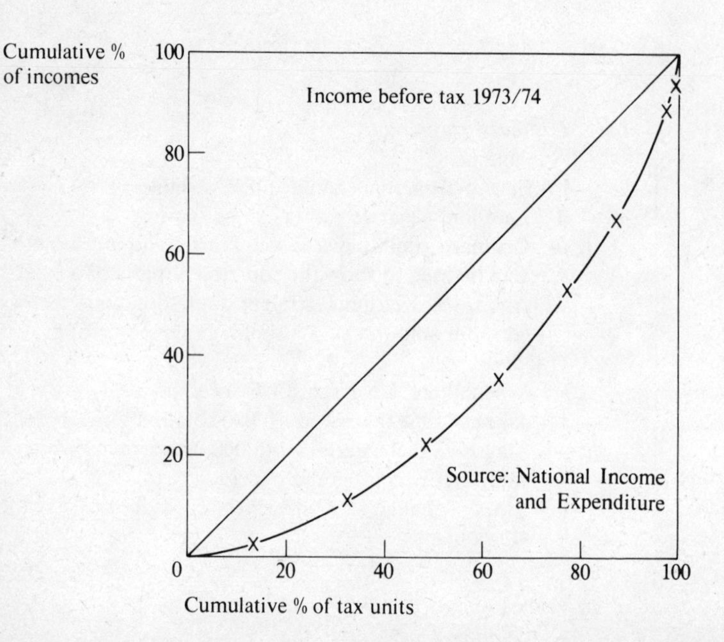

Figure A1.7

1.8

Month 1978	*Data*	*Cumulative data*	*Moving average total*
Jan.	60	60	759
Feb.	62	122	745
Mar.	69	191	730
Apr.	84	275	717
May	58	333	711
Jun.	40	373	704
Jul.	28	401	700
Aug.	45	446	689
Sep.	48	494	680
Oct.	46	540	666
Nov.	50	590	654
Dec.	52	642	642

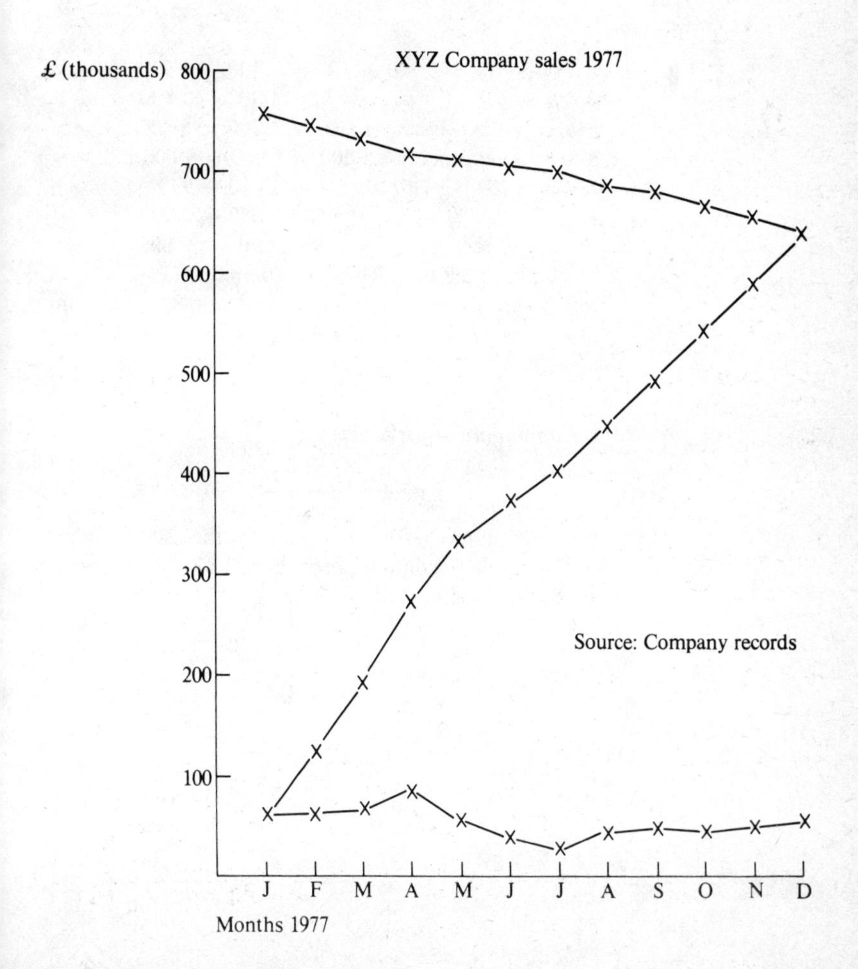

Figure A1.8

The Z chart displays to the businessman the picture of the sales for the past year. The Z chart consists of three elements: (i) the monthly sales data; (ii) the cumulative total; (iii) the moving average total. The moving average total, as it involves a total for the year, eliminates seasonal factors and gives a clear indication of whether sales are rising, falling or are constant. The cumulative total gives the cumulative sales up to the month in question and shows the progress so far that year. The plot of the data charts the actual sales pattern and displays any seasonal influences.

The moving average total is falling, thus sales are falling; the sales data appear to have a marked seasonality, with sales high in the spring, falling in the summer months and rising again in the autumn.

1.9 **a** 2 000 ± 5% = 2 000 ± 100 = 2 100 — 1 900
4 000 ± 300 = 4 300 — 3 600
Maximum expenses = 4 300 + 2 100 = 6 400.
Minimum expenses = 3 600 + 1 900 = 5 600

b 1 000 ± 70 = 1 070 — 930
8 ± 6.25% = 8 ± 0.5 = £8.50 — £7.50
Maximum receipts = 1 070 × £8.50 = £9 095.
Minimum receipts = 930 × £7.50 = £6 975

c Maximum profits = maximum receipts — minimum expenses = 9 095 — 5 600 = £3 495
Minimum profits = minimum receipts — maximum expenses = 6 975 — 6 400 = £575

d Maximum profit per article $= \frac{3\,495}{1\,070} =$ £3.27;

minimum $= \frac{575}{930} =$ £0.62

N.B. A common error is to divide by the smallest number 930 to obtain maximum profit, but £3 495 is calculated on the basis of selling 1 070 articles.

1.10

Range (£)	*Number*
40 - 44	1
45 - 49	3
50 - 54	3
55 - 59	6
60 - 64	8
65 - 69	9
70 - 74	7
75 - 79	5
80 - 84	4
85 - 89	2
Total	48

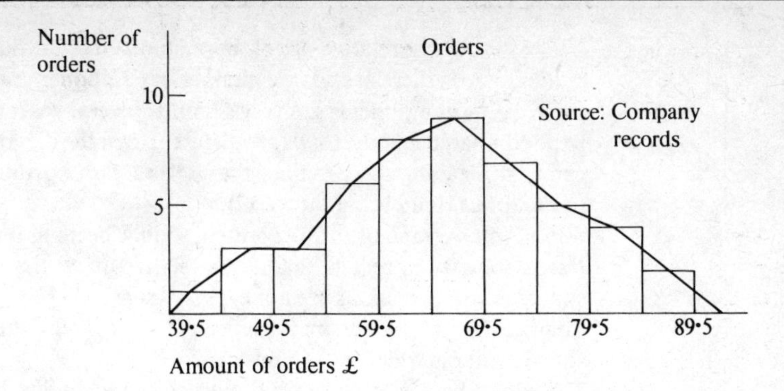

Figure A1.10

As values are recorded to the nearest £ the limits for the second interval are £44.50 to £49.49½.

1.11 a *Simple bar chart*. An uncomplicated diagram which is useful for showing data where it is important to concentrate on the size of the components. The main disadvantage is that it is not easy to visualise the relative size of the components in relation to the total.

Multiple bar chart. If we had data of U.K. Central Government receipts for say 1974, 1975 and 1976, the multiple bar chart would enable us to see how each component changes from year to year. As in the case of simple bar charts, it is not easy to visualise the relative size of the components in relation to the total. The diagram can become confusing if too many comparisons are attempted.

Component bar chart. Useful for showing how the total is split into components. A disadvantage is that the actual percentage that the component is of the total is not easy to see. This problem can be overcome by inserting the actual percentage on the component.

Percentage component bar chart. This diagram is useful if it is necessary to look at the relative size of the components. These can be readily noted from the scale. A disadvantage is that the totals are not given. If a comparison is made for two years and the total for one year is much greater, then the percentage component bar chart may show a reduction in a component when in absolute terms the component may have increased. In practice the choice between a component bar chart and a percentage component bar chart depends on what aspects of the data it is desirable to emphasise.

Pie chart. This is a popular method of displaying statistical data. It shows the relative size of the components. The difficulties arise if there are a large number of components and some of the components are very small. If several years are to be compared and the totals are different then the diagram can be misleading if the radii of the circles are proportional to the totals (see Figs. 1.9 to 1.12).

Pictogram. A popular form of presentation. The items illustrated are easy to recognise and the relative quantities are shown clearly; fractions are not easy to visualise. The main disadvantage of pictograms is the use of two or three dimensional symbols which aim to mislead.

b The choice here is between a pie chart and a percentage bar chart. The former is preferable. Did you follow all the instructions on page 6?

1.12 a Graph paper should be used. As the numbers range from 75 to 6 824, a logarithmic graph would be better. If logarithmic paper is not supplied or is not available from the invigilator ordinary graph paper should be used for this particular data.

b Two pie charts would be appropriate in this case; as the data are given as percentages, multiply by 3.6 to find the degrees.

c Percentage component bar chart is suitable for these data.

1.13 This frequency table has unequal class intervals; the first step therefore is to tackle this problem. There are a number of ways of doing this. One method is shown on page 8. In the solution supplied each percentage is based on a 5-year class interval.

0 – 4	13
5 – 9	8
10 – 14	8
15 – 19	8
20 – 24	9
25 – 44	8.75
45 – 59	4
60 – 64	3
65 – 74	2

The histogram shows a large number of under fives and relatively few people over the age of 45. (See Figure A1.13.)

1.14 a This part of the question is discussed in Exercise 1.6.

b As the quantities vary considerably, and probably comparative rates of growth need to be examined, logarithmic graph paper is probably the best method of presentation. However these data really need 4-cycle logarithmic paper and this

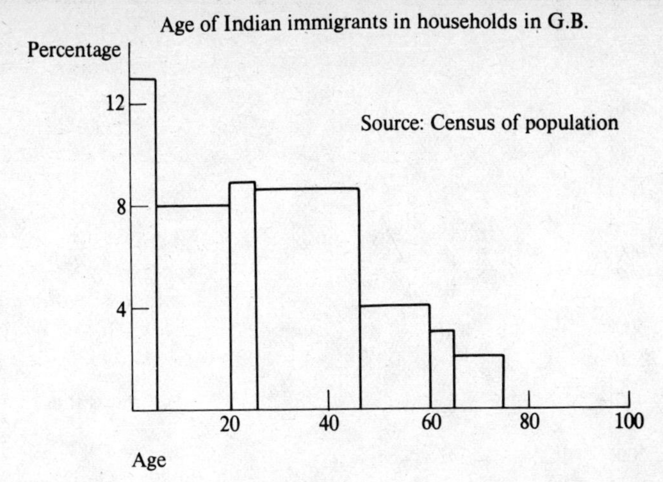

Figure A1.13

may not be available. In some cases it is possible to use one scale on the left-hand vertical axis, and a different scale on the right-hand axis.

The graphs have roughly similar upward slopes, so sales of cider in the 3 different containers have approximately the same rate of growth. (See Figure A1.14 overleaf.)

1.15

Cumulative no. of establishments	*Cumulative sales*	*% cum. estabs.*	*% cum. sales*
221	16	74.9	5.2
251	40	85.1	13.1
273	93	92.5	30.4
279	124	94.6	40.5
282	140	95.6	45.8
289	199	98.0	65.0
295	306	100.0	100.0

The Lorenz curve shows the concentration of the manufacture of electrical appliances primarily for domestic use is in a small number of large establishments. (See Figure A1.15 on page 215.)

1.16 **a** See 1.8

b

Sales	*Cumulative sales*	*Moving annual total*
46	46	580
44	90	591
52	142	602
57	199	610

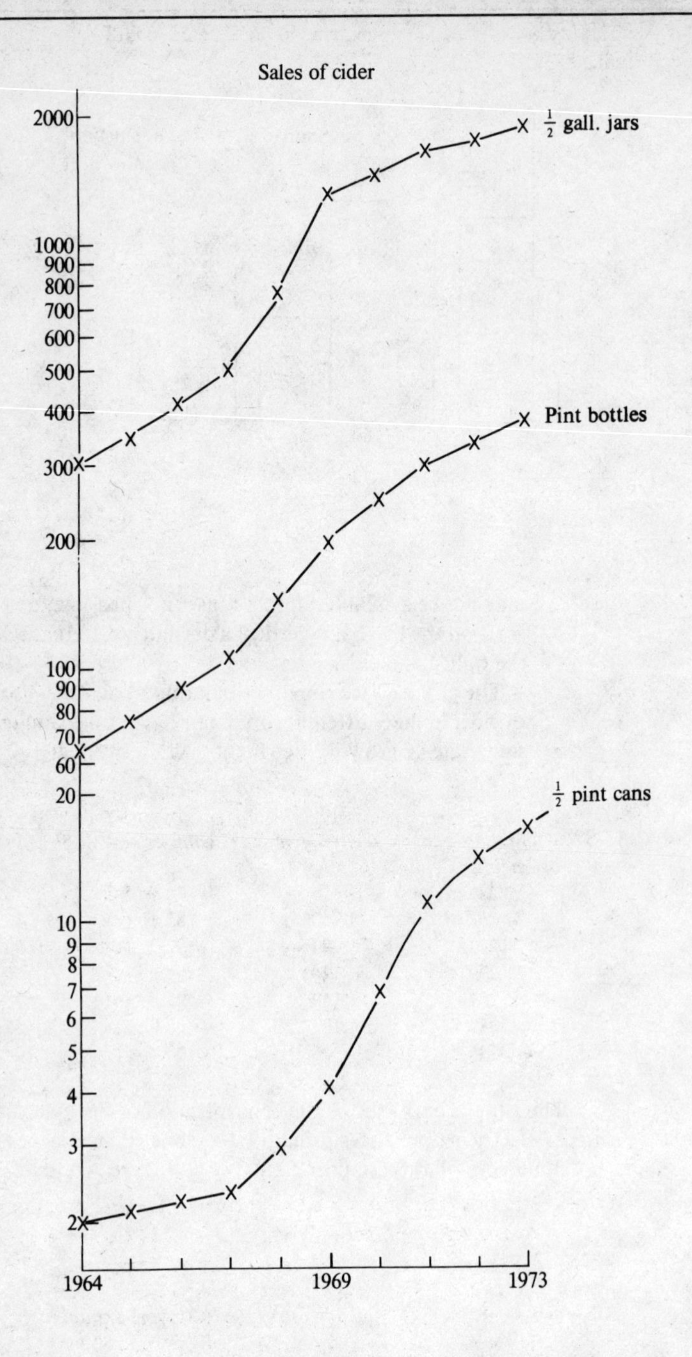
Sales of cider
2000
1000
900
800
700
600
500
400
300
200
100
90
80
70
60
20
10
9
8
7
6
5
4
3
2
½ gall. jars
Pint bottles
½ pint cans
1964
1969
1973

Figure A1.14

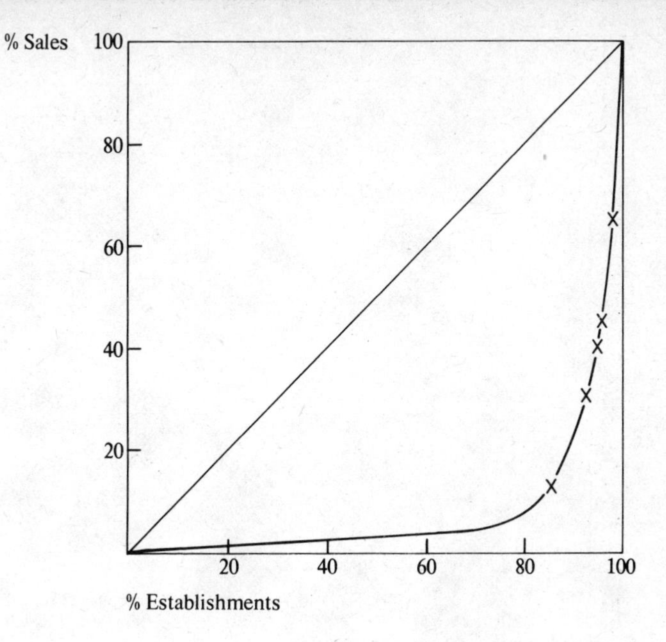

Figure A1.15 Source: *Census of Production 1970*

Sales	*Cumulative sales*	*Moving annual total*
62	261	616
70	331	623
83	414	635
57	471	639
46	517	643
39	556	645
35	591	644
60	651	651

c Sales rose steadily in 1972. The products of the company are affected by seasonal factors, peak sales being in the summer. (See Figure A1.16 overleaf.)

1.17 i Limits of the edge are $3.5 \pm 0.05 = 3.45$ to 3.55, thus limits of the perimeter are $3.45 \times 4 = 13.8$ and $3.55 \times 4 = 14.2$.

ii Limits of the volume of the block are $3.45 \times 3.45 \times 4.95 = 58.917$ and $3.55 \times 3.55 \times 5.05 = 63.643$.

Unbiased errors are errors which when combined tend to cancel out. For example in Exercise 1.10 the orders are recorded to the nearest £, with some orders above and some below the stated amount. In the total of all these orders, these errors (which are unbiased) would in general cancel out.

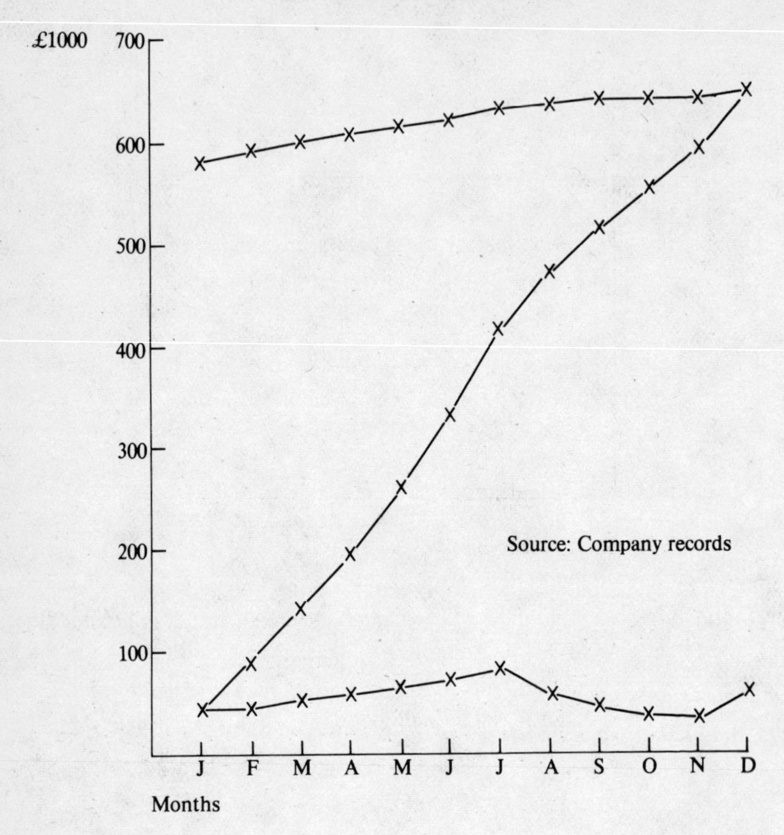

Figure A1.16

Source: *Company records*

(continued)

A biased error is one in which the errors are all in one direction. If in Exercise 1.10, instead of recording to the nearest £ the pence had been omitted, then in the total of all the orders these errors (which are biased) would cumulate. Biased errors thus cumulate and are therefore much more serious than unbiased errors.

Chapter 2

2.1

0 - 9	3	
10 - 19	1	
20 - 29	6	
30 - 39	13	
40 - 49	25	f_0
50 - 59	27	f_1
60 - 69	14	f_2
70 - 79	5	
80 - 89	6	

$L_1 = 50 \quad L_2 = 60$
$f_0 = 25 \quad f_1 = 27 \quad f_2 = 14$

$$\text{Mode} = L_1 + \left(\frac{f_1 - f_0}{2f_1 - f_0 - f_2}\right)(L_2 - L_1)$$
$$= 50 + \frac{(27 - 25)}{(2 \times 27 - 25 \div 14)}(60 - 50)$$
$$= 50 + \frac{2 \times 10}{15} = 50 + 1.33 = 51.33$$

Thus mode is 51.3 or 51 if marks are taken to the nearest whole number.

2.2 Arrange in order:
42, 58, 62, 69, 71, 73, 78, 80, 84, 92, 97, 103, 130, 170, 182
Median is 8th item = 80

2.2(A) Q_1 is 4th item = 69. Q_3 is 12th item = 103. Interquartile range is 103 – 69 = 34. Range is smallest item – largest item = 182 – 42 = 140. Quartile deviation is $\frac{103 - 69}{2} = \frac{34}{2} = 17$

2.3

Less than	*Cumulative frequency*
30	3
40	17
50	47
60	81
70	109
80	132
90	147
100	150

$\frac{n}{2} = \frac{150}{2} + 75. \frac{n}{4} = \frac{150}{4} = 37.5. \frac{3n}{4} = \frac{3 \times 150}{4} = 112.5$

$\frac{n}{10} = \frac{150}{10} = 15. \quad \frac{9n}{10} = \frac{9 \times 150}{10} = 135$

From Fig. 2.3(i), $M = 58$, $Q_1 = 47$, $Q_3 = 71$

$\frac{Q_3 - Q_1}{2} = \frac{71 - 47}{2} = 12$, $D_1 = 39$, $D_9 = 82$

From Fig. 2.3(ii) At 48 miles no. of vans = 39.

At 77 miles no. of vans = 126. Percentage = $\frac{126 - 39}{150} \times 100$

= 58%

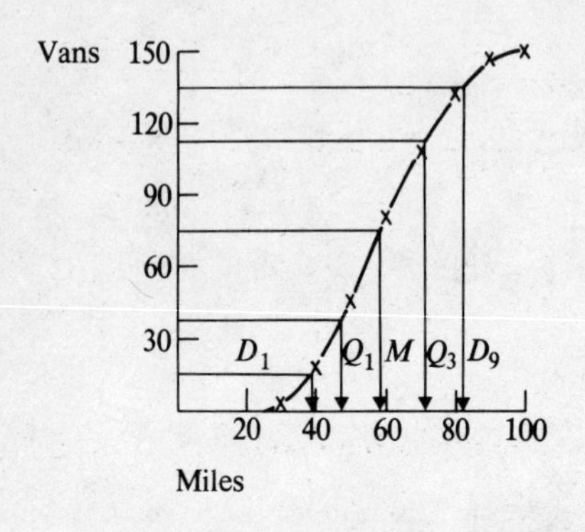

Figure A2.3 (i)

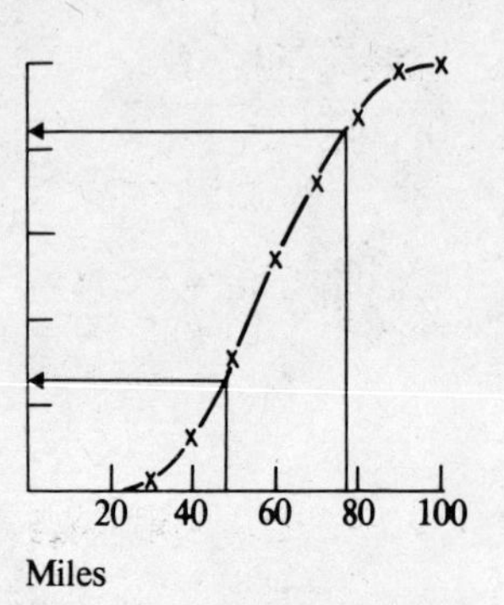

Figure A2.3 (ii)

2.3 **a**

Less than	*Cumulative frequency*	
30	3	⟵ D_1
40	17	⟵ Q_1
50	47	⟵ M
60	81	
70	109	
80	132	⟵ Q_3
90	147	⟵ D_9
100	150	

$\frac{n}{2} = \frac{150}{2} = 75.\ \frac{n}{4} = \frac{150}{4} = 37.5.\ \frac{3n}{4} = \frac{3 \times 150}{4}$

$= 112.5$

$L_1 = 50, \quad L_2 = 60 \qquad \Sigma f_1 = 47, \Sigma f_2 = 81$

$$M = L_1 + (L_2 - L_1)\left(\frac{\frac{n}{2} - \Sigma f_1}{\Sigma f_2 - \Sigma f_1}\right)$$

$$M = 50 + (60-50)\left(\frac{75 - 47}{81 - 47}\right) = 50 + \frac{10 \times 28}{34}$$

$= 50 + 8.24 = 58.24$ miles

$L_1 = 40, \quad L_2 = 50, \qquad \Sigma f_1 = 17, \quad \Sigma f_2 = 47$

$$Q_1 = L_1 + (L_2 - L_1)\left(\frac{\frac{n}{4} - \Sigma f_1}{\Sigma f_2 - \Sigma f_1}\right)$$

$$Q_1 = 40 + (50-40)\left(\frac{37.5 - 17}{47 - 17}\right) = 40 + \frac{10 \times 20.5}{30}$$

$= 40 + 6.83 = 46.83$ miles

$L_1 = 70, \quad L_2 = 80, \qquad \Sigma f_1 = 109, \quad \Sigma f_2 = 132$

$$Q_3 = L_1 + (L_2 - L_1)\left(\frac{\frac{3n}{4} - \Sigma f_1}{\Sigma f_2 - \Sigma f_1}\right)$$

$$Q_3 = 70 + (80 - 70)\left(\frac{112.5 - 109}{132 - 109}\right) = 70 + \frac{10 \times 3.5}{23}$$

$$= 70 + 1.52 = 71.52 \text{ miles}$$

$L_1 = 30, \quad L_2 = 40, \qquad \Sigma f_1 = 3, \quad \Sigma f_2 = 17$

$$D_1 = L_1 + (L_2 - L_1)\left(\frac{\frac{n}{10} - \Sigma f_1}{\Sigma f_2 - \Sigma f_1}\right)$$

$$D_1 = 30 + (40 - 30)\left(\frac{15 - 3}{17 - 3}\right) = 30 + \frac{10 \times 12}{14}$$

$$= 30 + 8.57 = 38.57 \text{ miles}$$

$L_1 = 80, \quad L_2 = 90 \qquad \Sigma f_1 = 132, \quad \Sigma f_2 = 147$

$$D_9 = L_1 + (L_2 - L_1)\left(\frac{\frac{9n}{10} - \Sigma f_1}{\Sigma f_2 - \Sigma f_1}\right)$$

$$D_9 = 80 + (90 - 80)\left(\frac{135 - 132}{147 - 132}\right) = 80 + \frac{10 \times 3}{15}$$

$$= 80 + 2 = 82 \text{ miles}$$

$$\text{Quartile deviation} = \frac{Q_3 - Q_1}{2} = \frac{71.52 - 46.83}{2}$$

$$= \frac{24.69}{2} = 12.35 \text{ miles}$$

2.4

Less than	*Cumulative frequency*
101	13.4
501	65.1
1 001	84.3
2 001	93.8
5 001	97.8
10 001	98.7
50 001	99.6
100 001	100.0

$$\frac{n}{2} = \frac{100}{2} = 50 \qquad \frac{n}{4} = \frac{100}{4} = 25 \qquad \frac{3n}{4} = \frac{3 \times 100}{4} = 75$$

From graph: M = 340 shares
Q_1 = 160 shares
Q_3 = 670 shares

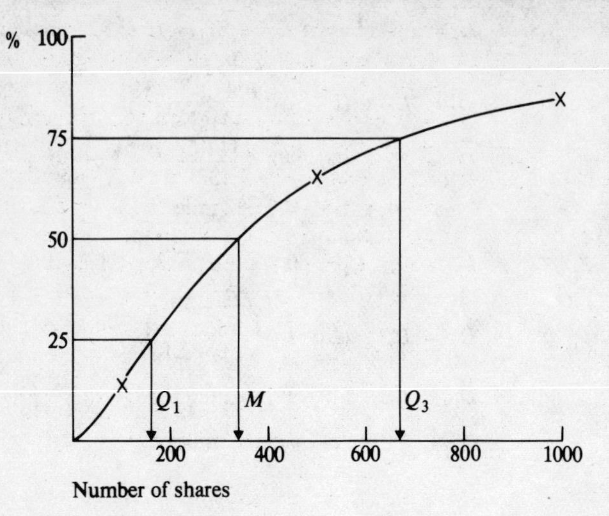

Figure A2.4

2.5 **a** *Mean* — Advantages

i Is well known — in common usage it is the 'average'.
ii A fairly simple calculation.
iii Takes into account all the observations.
iv The most commonly used measure of location.
v Suitable for algebraic manipulation.
vi Used in statistical theory.

Disadvantages

i Can be distorted by a few very high or very low values.
ii The mean may not correspond to an actual value, e.g. mean number of children 2.4.
iii Unlike the median and mode, cannot be found graphically.
iv In distributions with open-ended classes, the value of the mean depends on the sometimes very arbitrary choice of central values.

Median — Advantages

i Not distorted by a few very high or very low values.
ii Not affected by open-ended classes.
iii Do not need to know all the values to compute the median; we do not need to know the salary of the Managing Director to find median salary.
iv The middle value is a particularly useful measure of average in skew distributions, hence its use in income distributions.
v In part (b) of this question the median is an actual value; the mean need not be.

Disadvantages

i The median is not suitable for mathematical or theoretical work.

ii If extreme values need to be taken into account, median unsuitable.

iii When data are in the form of a frequency table with class intervals the estimation of the median is arbitrary — the graphical method assumes a curve and the calculation method assumes a straight line; the results can often be markedly different. In Exercise 2.4 calculation method gives 384 shares, compared with 340 by graphical method.

Mode — Advantages

i Not distorted by a few very high or very low values.

ii The mode is an actual value, e.g. for number of children in a family mode is 2.

iii It is easily understood.

iv Do not need to know all the values.

Disadvantages

i Unsuitable for mathematical or theoretical work.

ii When a distribution has two or more peaks there is more than one mode.

iii Cannot be found with precision when data is grouped — method used makes some arbitary assumptions.

b 28 + 9 + 14 + 19 + 3 + 33 + 22 + 7 + 14 + 26 + 10 + 16 + 20 = 221

$$\bar{x} = \frac{\Sigma x}{n} = \frac{221}{13} = 17$$

3, 7, 9, 10, 14, 14, 16, 19, 20, 22, 26, 28, 33

Median is middle value, i.e. 7th value = 16

Mode is the observation that occurs most often = 14

2.6. Mean is the 'average' and is equal to the sum of the observations divided by the number of observations.

Median is the value such that half the observations exceed the median and half are less than the median. If the observations are ranked in order of size, the median is the middle observation.

Mode is the observation which occurs most often.

Range is the difference between the largest and smallest observations.

15 + 14 + 17 + ... + 16 + 20 = 256.

$$\text{Mean} = \frac{\Sigma x}{n} = \frac{256}{16} = 16$$

Put in order:

11, 13, 14, 15, 15, 15, 16, 16, 16, 16, 17, 17, 18, 18, 19, 20

Middle value is midway between 8th and 9th observations. The 8th and 9th observations are both 16, hence median is 16.

The observation that occurs most often is 16, thus mode = 16.
Range = largest observation – smallest observation = 20 – 11 = 9.

2.7 **a**

Range	f
50 – 59	1
60 – 69	2
70 – 79	8
80 – 89	18 = f_0
90 – 99	23 = f_1
100 – 109	21 = f_2
110 – 119	15
120 – 129	9
130 – 139	3

$$L_1 = 90, L_2 = 100$$
$$f_0 = 18, f_1 = 23, f_2 = 21$$
$$\text{Mode} = L_1 + \left(\frac{f_1 - f_0}{2f_1 - f_0 - f_2}\right)(L_2 - L_1)$$
$$= 90 + \left(\frac{23 - 18}{2 \times 23 - 18 - 21}\right)(100 - 90)$$
$$= 90 + \frac{5 \times 10}{7} = 90 + 7.14 = 97.14$$

b

Less than	*Cumulative frequency*	
60	1	
70	3	
80	11	
90	29	← M
100	52	
110	73	
120	88	
130	97	
140	100	

$$\frac{n}{2} = \frac{100}{2} = 50, \frac{n}{4} = \frac{100}{4} = 25, \frac{3n}{4} = \frac{3 \times 100}{4} = 75$$
$$L_1 = 90, L_2 = 100$$
$$\Sigma f_1 = 29, \Sigma f_2 = 52$$
$$M = L_1 + (L_2 - L_1)\left(\frac{\frac{n}{2} - \Sigma f_1}{\Sigma f_2 - \Sigma f_1}\right)$$
$$M = 90 + (100 - 90)\left(\frac{50 - 29}{52 - 29}\right) = 90 + \frac{10 \times 21}{23}$$
$$= 90 + 9.13 = 99.13$$

2.8 *Tankers*

Less than	*Cumulative frequency*	
5	137	
10	250	← M
15	401	← Q_3
20	463	
25	478	
30	486	

$$\frac{n}{2} = \frac{486}{2} = 243 \qquad \frac{3n}{4} = \frac{3 \times 486}{4} = 364.5$$

$L_1 = 5, L_2 = 10$
$\Sigma f_1 = 137, \Sigma f_2 = 250$

$$M = L_1 + (L_2 - L_1)\left(\frac{\frac{n}{2} - \Sigma f_1}{\Sigma f_2 - \Sigma f_1}\right)$$

$$M = 5 + (10 - 5)\left(\frac{243 - 137}{250 - 137}\right) = 5 + \frac{5 \times 106}{113}$$

$$= 5 + 4.69 = 9.69 \text{ years}$$

$L_1 = 10, \quad L_2 = 15, \quad \Sigma f_1 = 250, \quad \Sigma f_2 = 401.$

$$Q_3 = L_1 + (L_2 - L_1)\left(\frac{\frac{3n}{4} - \Sigma f_1}{\Sigma f_2 - \Sigma f_1}\right)$$

$$Q_3 = 10 + (15 - 10)\left(\frac{364.5 - 250}{401 - 250}\right) = 10 + \frac{5 \times 114.5}{151}$$

$$= 10 + 3.79 = 13.79 \text{ years}$$

Passenger vessels

Less than	*Cumulative frequency*	
5	18	
10	34	
15	61	← M
20	81	← Q_3
25	112	
30	129	

$$\frac{n}{2} = \frac{129}{2} = 64.5 \qquad \frac{3n}{4} = \frac{3 \times 129}{4} = 96.75$$

$L_1 = 15, L_2 = 20$
$\Sigma f_1 = 61, \Sigma f_2 = 81$

$$\text{M} = L_1 + (L_2 - L_1)\left(\frac{\frac{n}{2} - \Sigma f_1}{\Sigma f_2 - \Sigma f_1}\right)$$

$$M = 15 + (20 - 15)\left(\frac{64.5 - 61}{81 - 61}\right) = 15 + \frac{5 \times 3.5}{20}$$

$$= 15 + 0.88$$
$$= 15.88 \text{ years}$$

$L_1 = 20, \quad L_2 = 25 \qquad \Sigma f_1 = 81, \quad \Sigma f_2 = 112$

$$Q_3 = L_1 + (L_2 - L_1)\,\frac{\frac{3n}{4} - \Sigma f_1}{\Sigma f_2 - \Sigma f_1}$$

$$Q_3 = 20 + (25 - 20)\,\frac{96.75 - 81}{112 - 81}$$

$$= 20 + \frac{5 \times 15.75}{31} = 20 + 2.54 = 22.54 \text{ years}$$

2.9

Less than	*Cumulative frequency* *1971*	*1974*
20	118	64
25	840	758
30	1128	1444
35	1240	1607
50	1352	1807
80	1529	1972

Ignore those whose age not stated.

Position	*1971*	*Value*	*1974 Position*	*Value*
$\frac{n}{2}$	$= \frac{1\,529}{2} = 764.5$	24	$\frac{1\,972}{2} = 986$	26
$\frac{n}{4}$	$= \frac{1\,529}{4} = 382.25$	21	$\frac{1\,972}{4} = 493$	24
$\frac{3n}{4}$	$= \frac{3 \times 1\,529}{4} = 1\,146.75$	30	$\frac{3 \times 1\,972}{4} = 1\,479$	32
$\frac{Q_3 - Q_1}{2}$		4.5		4

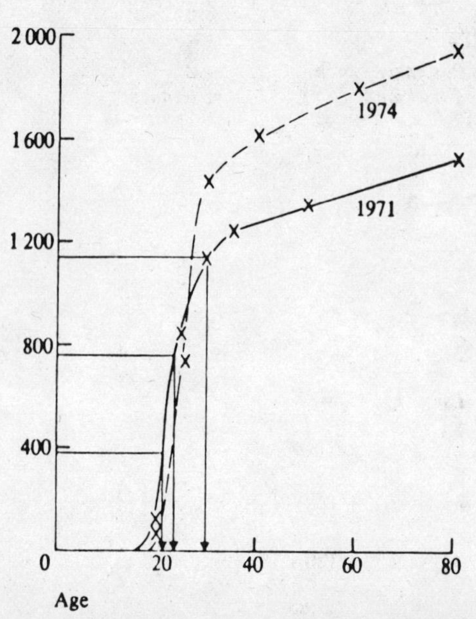

Figure A2.9 Construction lines for 1974 not drawn.

2.10 a With income distributions the mean usually exceeds the median because a small number of persons with high incomes push the mean upwards. It is argued that the 'average' man is better represented by the median income — the income of the person in the middle of the range. The median is used by union negotiators; the published statistics of the Department of Employment give prominence to median incomes.

The lower quartile and the upper quartile are also given in the Department of Employment income statistics. The lower quartile income shows the income level which is exceeded by $\frac{3}{4}$ of the employees; the upper quartile income shows the income level which is exceeded by $\frac{1}{4}$ of the employees. The quartiles give values which have clear meaning. In the case of income distributions the upper quartile is usually farther from the median than the lower quartile, if the standard deviation (see Chapter 3) is used the asymmetric nature of the distribution is not shown.

b

Less than	*Cumulative frequency*	
10	2	
20	17	← Q_1
30	27	
40	36	
50	47	← M
60	59	
70	70	← Q_3
80	78	
100	89	
120	100	

$$\frac{n}{2} = \frac{100}{2} = 50 \qquad \frac{n}{4} = \frac{100}{4} = 25$$

$$\frac{3n}{4} = \frac{3 \times 100}{4} = 75$$

$$L_1 = 50, \quad L_2 = 60$$
$$\Sigma f_1 = 47, \quad \Sigma f_2 = 59$$

$$M = L_1 + (L_2 - L_1)\,\frac{\frac{n}{2} - \Sigma f_1}{\Sigma f_2 - \Sigma f_1}$$

$$M = 50 + (60 - 50)\,\frac{50 - 47}{59 - 47} = 50 + \frac{10 \times 3}{12}$$

$$= 50 + 2.5 = £52.50$$

$$L_1 = 20, \quad L_2 = 30, \qquad \Sigma f_1 = 17, \quad \Sigma f_2 = 27$$

$$Q_1 = L_l + (L_2 - L_1)\,\frac{\frac{n}{4} - \Sigma f_1}{\Sigma f_2 - \Sigma f_1}$$

$$Q_1 = 20 + (30 - 20)\,\frac{25 - 17}{27 - 17} = 20 + \frac{10 \times 8}{10}$$

$$= 20 + 8 = £28$$

$L_1 = 70, \quad L_2 = 80, \qquad \Sigma f_1 = 70, \quad \Sigma f_2 = 78$

$$Q_3 = L_1 + (L_2 - L_1) \frac{\frac{3n}{4} - \Sigma f_1}{\Sigma f_2 - \Sigma f_1}$$

$$Q_3 = 70 + (80 - 70) \frac{75 - 70}{78 - 70} = 70 + \frac{10 \times 5}{8}$$

$$= 70 + 6.25 = £76.25$$

$L_1 = 100, \quad L_2 = 120 \qquad \Sigma f_1 = 89, \quad \Sigma f_2 = 100$

$$D_9 = L_1 + (L_1 - L_1) \frac{\frac{9n}{10} - \Sigma f_1}{\Sigma f_2 - \Sigma f_1}$$

$$D_9 = 100 + (120 - 100) \frac{90 - 89}{100 - 89} = 100 + \frac{20 \times 1}{11}$$

$$= 100 + 1.82 = £101.82$$

$$\text{Quartile deviation} = \frac{Q_3 - Q_1}{2} = \frac{76.25 - 28}{2}$$

$$= £24.13$$

	Greater London	*Scotland*
Q_1	£28.00	£27.30
M	£52.50	£46.20
Q_3	£76.25	£68.20
D_9	£101.82	£95.00
Q.D.	£24.13	£20.50

Incomes in Greater London, lower quartile apart, are markedly higher than in Scotland. Wages have always been higher in London, and many employees are paid a 'London' allowance to compensate for the higher cost of living in London.

2.11 a Median is that value such that half the observations exceed the median and half are less than the median.

b The lower quartile is that value such that three quarters of the observations exceed the lower quartile and one quarter are less than the lower quartile. The upper quartile is such that one quarter of the observations are greater than the upper quartile and three quarters are less than the upper quartile.

c The quartile deviation is the average deviation of the quartiles from the median and is equal to $\frac{Q_3 - Q_1}{2}$

0, 0, 1, 2, 3, 4, 4, 5, 6, 7, 7, 8, 9, 9, 10
Median is 8th item = 5. Q_1 is 4th item = 2. Q_3 is 12th item = 8.

$$\text{Quartile deviation} = \frac{Q_3 - Q_1}{2} = \frac{8 - 2}{2} = 3$$

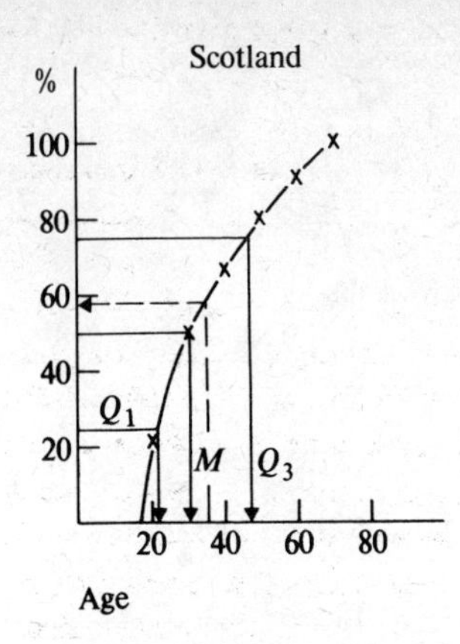

Figure A2.12 (A)

East Anglia

Figure A2.12 (B)

2.12 This data is already given in a cumulative form.

$$\frac{n}{2} = \frac{100}{2} = 50 \qquad \frac{n}{4} = \frac{100}{4} = 25 \qquad \frac{3n}{4} = \frac{3 \times 100}{4} = 75$$

		Scotland	*East Anglia*
Reading from the graph	M	30	33
	Q_1	20.5	22
	Q_3	46.5	54
	Q.D.	13	16
More than mean:		42%	43%

2.13

Less than	*Cumulative frequency*	
2	1045	
5	2246	← Q_1
8	3278	
13	4530	← M
26	5564	← Q_3
52	6221	
104	7183	

$$\frac{n}{2} = \frac{7\,183}{2} = 3\,591.5 \qquad \frac{n}{4} = \frac{7\,183}{4} = 1\,795.75$$

$$\frac{3n}{4} = \frac{3 \times 7\,183}{4} = 5\,387.25$$

$$L_1 = 8, \quad L_2 = 13, \qquad \Sigma f_1 = 3\,278, \quad \Sigma f_2 = 4\,530$$

$$M = L_1 + (L_2 - L_1)\left(\frac{\frac{n}{2} - \Sigma f_1}{\Sigma f_2 - \Sigma f_1}\right)$$

$$M = 8 + (13 - 8)\left(\frac{3\,591.5 - 3\,278}{4\,530 - 3\,278}\right) = 8 + \frac{5 \times 313.5}{1\,252}$$

$$= 8 + 1.25 = 9.25$$

$L_1 = 2, \quad L_2 = 5, \qquad \Sigma f_1 = 1\,045, \quad \Sigma f_2 = 2\,246$

$$Q_1 = L_1 + (L_2 - L_1)\left(\frac{\frac{n}{4} - \Sigma f_1}{\Sigma f_2 - \Sigma f_1}\right)$$

$$Q_1 = 2 + (5 - 2)\left(\frac{1\,795.75 - 1\,045}{2\,246 - 1\,045}\right) = 2 + \frac{3 \times 750.75}{1\,201}$$

$$= 2 + 1.88 = 3.88 \text{ weeks}$$

$L_1 = 13, \quad L_2 = 26, \qquad \Sigma f_1 = 4\,530, \quad \Sigma f_2 = 5\,564$

$$Q_3 = L_1 + (L_2 - L_1)\left(\frac{\frac{3n}{4} - \Sigma f_1}{\Sigma f_2 - \Sigma f_1}\right)$$

$$Q_3 = 13 + (26 - 13)\left(\frac{5\,387.25 - 4\,530}{5\,564 - 4\,530}\right)$$

$$= 13 + \frac{13 \times 857.25}{1\,034} = 13 + 10.78 = 23.78 \text{ weeks}$$

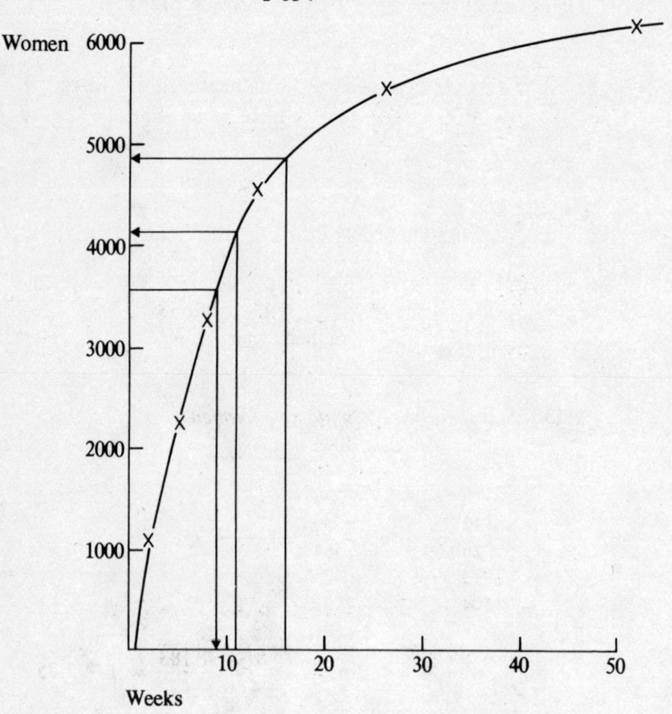

Figure A2.13

$$\text{Quartile deviation} = \frac{Q_3 - Q_1}{2} = \frac{23.78 - 3.88}{2}$$

$$= \frac{19.9}{2} = 9.95$$

Reading from the graph, median is 9 weeks.

Cumulative number of unemployed women at 16 weeks	4 860
Cumulative number of unemployed women at 11 weeks	4 120
Number of women unemployed between 11 and 16 weeks	740

Chapter 3

3.1 $\Sigma x = 41 + 72 + 31 + 84 + 54 + 52 + 60 + 49 + 43 + 64 = 550$

$$\frac{\Sigma x}{n} = \frac{550}{10} = 55$$

3.2

x	f	fx
1	1	1
2	2	4
3	8	16
4	21	84
5	33	165
6	26	156
7	5	35
9	4	36
	100	505

$$\overline{x} = \frac{\Sigma fx}{\Sigma f} = \frac{505}{100} = 5.05 \text{ rooms}$$

3.3

x	f	fx
5	3	15
15	8	120
25	16	400
35	13	455
45	7	315
55	3	165
	50	1 470

$$\overline{x} = \frac{\Sigma fx}{\Sigma f} = \frac{1\,470}{50} = 29.4 = \text{£}29\,400 \text{ (since units £1 000)}$$

3.4 Mean = 55.

Arithmetic differences are 14, 17, 24, 29, 1, 3, 5, 6, 12, 9

$$\text{Mean deviation} = \frac{\Sigma|x - \overline{x}|}{n}$$

$$= \frac{14 + 17 + 24 + 29 + 1 + 3 + 5 + 6 + 12 + 9}{10}$$

$$= \frac{120}{10} = 12$$

3.5

x	f	fx	fx^2
1	1	1	1
2	2	4	8
3	8	24	72
4	21	84	336

x	f	fx	fx^2
5	33	165	825
6	26	156	936
7	5	35	245
9	4	36	324
	100	505	2 747

$$\sigma = \sqrt{\frac{\Sigma fx^2}{\Sigma f} - \left(\frac{\Sigma fx}{\Sigma f}\right)^2} = \sqrt{\frac{2\,747}{100} - \left(\frac{505}{100}\right)^2}$$
$$= \sqrt{27.47 - 25.502\,5}$$
$$= \sqrt{1.967\,5} = 1.403 \text{ rooms}$$

3.6

x	f	fx	fx^2
5	3	15	75
15	8	120	1 800
25	16	400	10 000
35	13	455	15 925
45	7	315	14 175
55	3	165	9 075
	50	1 470	51 050

$$\sigma = \sqrt{\frac{\Sigma fx^2}{\Sigma f} - \left(\frac{\Sigma fx}{\Sigma f}\right)^2} = \sqrt{\frac{51\,050}{50} - \left(\frac{1\,470}{50}\right)^2}$$
$$= \sqrt{1\,021 - 864.36} = \sqrt{156.64} = 12.52 = \pounds 12\,520$$

3.7 i Method not using an assumed mean:

x	f	fx	fx^2
250	5	1 250	312 500
750	12	9 000	6 750 000
1 250	18	22 500	28 125 000
1 750	23	40 250	70 437 500
2 250	20	45 000	101 250 000
2 750	11	30 250	83 187 500
3 250	11	35 750	116 187 500
	100	184 000	406 250 000

$$\overline{x} = \frac{\Sigma fx}{\Sigma f} = \frac{184\,000}{100} = \pounds 1\,840$$
$$\sigma = \sqrt{\frac{\Sigma fx^2}{\Sigma f} - \left(\frac{\Sigma fx}{\Sigma f}\right)^2} = \sqrt{\frac{406\,250\,000}{100} - \left(\frac{184\,000}{100}\right)^2}$$
$$= \sqrt{4\,062\,500 - 3\,385\,600} = \sqrt{676\,900} = 822.74$$
$$= \pounds 823$$

ii Method using an assumed mean.
Let assumed mean = 1 750, class interval = 500.

x	$x' = \dfrac{x - 1\,750}{500}$	f	fx'	fx'^2
250	−3	5	−15	45
750	−2	12	−24	48
1 250	−1	18	−18	18
1 750	0	23	0	0
2 250	+1	20	+20	20
2 750	+2	11	+22	44
3 250	+3	11	+33	99
		100	+18	274

$A = 1\,750 \quad C = 500$

$$\overline{x} = A + \frac{C\Sigma fx'}{\Sigma f} = 1\,750 + 500 \times \frac{18}{100}$$

$$= 1750 + 90 \quad = £1840$$

$$\sigma = C\sqrt{\frac{\Sigma fx'^2}{\Sigma f} - \left(\frac{\Sigma fx'}{\Sigma f}\right)^2}$$

$$= 500\sqrt{\frac{274}{100} - \left(\frac{18}{100}\right)^2}$$

$$= 500\sqrt{2.74 - .0324} = 500\sqrt{2.7076}$$

$$= 500 \times 1.6455 = £823$$

i Method not using an assumed mean

x	f	fx	fx^2
250	3	750	187 500
750	10	7 500	5 625 000
1 250	32	40 000	50 000 000
1 750	30	52 500	91 875 000
2 250	15	33 750	75 937 500
2 750	5	13 750	37 812 500
3 250	5	16 250	52 812 500
	100	164 500	314 250 000

$$\overline{x} = \frac{\Sigma fx}{\Sigma f} = \frac{164\,500}{100} = £1\,645$$

$$\sigma = \sqrt{\frac{\Sigma fx^2}{\Sigma f} - \left(\frac{\Sigma fx}{\Sigma f}\right)^2} = \sqrt{\frac{314\,250\,000}{100} - \left(\frac{164\,500}{100}\right)^2}$$

$$= \sqrt{3\,142\,500 - 2\,706\,025} = \sqrt{436\,475} = 660.66$$

$$= £661$$

ii Method using an assumed mean.
Let assumed mean = 1 750, class interval = 500

x	$x' = \dfrac{x - 1\,750}{500}$	f	fx'	fx'^2
250	−3	3	−9	27
750	−2	10	−20	40
1 250	−1	32	−32	32

x	$x' = \frac{x - 1\ 750}{500}$	f	fx'	fx'^2
1 750	0	30	0	0
2 250	+1	15	+15	15
2 750	+2	5	+10	20
3 250	+3	5	+15	45
		100	−21	179

$A = 1\ 750 \quad C = 500$

$$\bar{x} = A + \frac{C\Sigma fx'}{\Sigma f} = 1\ 750 + 500 \times \frac{-21}{100}$$

$$= 1\ 750 + (-105) = £1\ 645$$

$$\sigma = C\sqrt{\frac{\Sigma fx'^2}{\Sigma f} - \left(\frac{\Sigma fx'}{\Sigma f}\right)^2}$$

$$= 500\sqrt{\frac{179}{100} - \left(\frac{-21}{100}\right)^2}$$

$$= 500\sqrt{1.79 - 0.044\ 1} = 500\sqrt{1.745\ 9}$$

$$= 500 \times 1.321\ 3 = £661$$

3.8 In 3.4 we found $|x - \bar{x}|$. We need $\Sigma(x - \bar{x})^2$. Square each $|x - \bar{x}|$ and add. $\Sigma(x - \bar{x})^2 = 196 + 289 + 576 + 841 + 1 + 9 + 25 + 36 + 144 + 81 = 2\ 198$

$$\sigma = \sqrt{\frac{\Sigma(x - \bar{x})^2}{n}} = \sqrt{\frac{2\ 198}{10}} = \sqrt{219.8} = 14.83.$$

$$\sigma = \sqrt{\frac{\Sigma(x - \bar{x})^2}{n - 1}} = \sqrt{\frac{2\ 198}{9}} = \sqrt{244.22} = 15.63.$$

3.9 **i** $\Sigma x = 22 + 18 + 21 + 24 + 19 + 20 + 21 + 17 + 25 + 19 + 23 + 22 + 23 + 21 = 315$

$$\bar{x} = \frac{\Sigma x}{n} = \frac{315}{15} = 21$$

ii $|x - \bar{x}| = 1 + 3 + 0 + 3 + 2 + 1 + 0 + 4 + 4 + 2 + 2 + 1 + 1 + 2 + 0 = 26$

$$\text{Mean deviation} = \frac{\Sigma|x - \bar{x}|}{n} = \frac{26}{15} = 1.73$$

iii $\Sigma(x - \bar{x})^2 = 1 + 9 + 9 + 4 + 1 + 16 + 16 + 4 + 4 + 1 + 1 + 4 = 70$

$$\sigma = \sqrt{\frac{\Sigma(x - \bar{x})^2}{n}} = \sqrt{\frac{70}{15}} = \sqrt{4.666\ 7} = 2.16$$

$$\sigma = \sqrt{\frac{\Sigma(x - \bar{x})^2}{n - 1}} = \sqrt{\frac{70}{14}} = \sqrt{5} = 2.24$$

3.10 *Firm A*:

x	f	fx
45	6	270
55	12	660
65	8	520
75	5	375
85	14	1 190
95	3	285
	48	3 300

Firm B:

x	f	fx
45	11	495
55	15	825
65	16	1 040
75	9	675
85	6	510
95	2	190
	59	3 735

$$\bar{x} = \frac{\Sigma fx}{\Sigma f} = \frac{3\ 300}{48} = £68.75 \quad \bar{x} = \frac{\Sigma fx}{\Sigma f} = \frac{3\ 735}{59} = £63.31$$

3.11 *1949 or earlier*

x	f	fx	fx^2
0	3	0	0
1	20	20	20
2	31	62	124
3	19	57	171
4	13	52	208
5	6	30	150
7	8	56	392
	100	277	1 065

$$\bar{x} = \frac{\Sigma fx}{\Sigma f} = \frac{277}{100} = 2.77 \text{ children}$$

$$\sigma = \sqrt{\frac{\Sigma fx^2}{\Sigma f} - \left(\frac{\Sigma fx}{\Sigma f}\right)^2} = \sqrt{\frac{1\ 065}{100} - \left(\frac{277}{100}\right)^2}$$

$$= \sqrt{10.65 - 7.672\ 9} = \sqrt{2.977\ 1}$$

$$= 1.73 \text{ children}$$

1955 – 1959

x	f	fx	fx^2
0	6	0	0
1	11	11	11
2	37	74	148
3	28	84	252
4	11	44	176
5	5	25	125
7	2	14	98
	100	252	810

$$\bar{x} = \frac{\Sigma fx}{\Sigma f} = \frac{252}{100} = 2.52 \text{ children}$$

$$\sigma = \sqrt{\frac{\Sigma fx^2}{\Sigma f} - \left(\frac{\Sigma fx}{\Sigma f}\right)^2} = \sqrt{\frac{810}{100} - \left(\frac{252}{100}\right)^2}$$

$$= \sqrt{8.1 - 6.350\ 4} = \sqrt{1.749\ 6} = 1.32 \text{ children}$$

3.12 a i The mean or arithmetic mean is the same as the word average in normal usage. For example, the average or the mean number of goals scored by a team in the last 10 football matches is obtained by finding the total number of goals scored in the 10 matches and dividing by 10.

ii The mean as explained above gives the average. We are often interested in obtaining a measure which shows how near or how far the actual observed values are to the mean. The standard deviation is such a measure. Using again the example on goals scored, if the standard deviation is small it would mean that the team is consistent, i.e. scores nearly the same number of goals each week. If the standard deviation were relatively large this would mean that the team were inconsistent — the number of goals would vary about the mean, sometimes scoring no goals and sometimes scoring plenty of goals.

We calculate the standard deviation by finding the mean number of goals scored; for each match we find the difference between the actual number of goals scored and the mean. We square this difference and total these squares for each of the ten matches, divide this total by 10 and then take the square root of the resulting figure.

b

x	f	fx	fx^2
60	6	360	21 600
100	6	600	60 000
140	12	1 680	235 200
180	12	2 160	388 800
220	11	2 420	532 400
260	14	3 640	946 400
300	11	3 300	990 000
340	8	2 720	924 800
380	20	7 600	2 888 000
	100	24 480	6 987 200

$$\bar{x} = \frac{\Sigma fx}{\Sigma f} = \frac{24\,480}{100} = £244.80$$

$$\sigma = \sqrt{\frac{\Sigma fx^2}{\Sigma f} - \left(\frac{\Sigma fx}{\Sigma f}\right)^2} = \sqrt{\frac{6\,987\,200}{100} - \left(\frac{24\,480}{100}\right)^2}$$

$$= \sqrt{69\,872 - 59\,927.04} = \sqrt{9\,944.96}$$

$$= £99.72$$

3.13

x	f	fx	fx^2
5	5	25	125
15	10	150	2 250
25	25	625	15 625
35	30	1 050	36 750
50	17	850	42 500
70	11	770	53 900
90	2	180	16 200
	100	3 650	167 350

$$\bar{x} = \frac{\Sigma fx}{\Sigma f} = \frac{3\ 650}{100} = 36.5 \text{ minutes}$$

$$\sigma = \sqrt{\frac{\Sigma fx^2}{\Sigma f} - \left(\frac{\Sigma fx}{\Sigma f}\right)^2} = \sqrt{\frac{167\ 350}{100} - \left(\frac{3\ 650}{100}\right)^2}$$
$$= \sqrt{1\ 673.50 - 1\ 332.25} = \sqrt{341.25}$$
$$= 18.5 \text{ minutes}$$

3.14 As the numbers are large, this exercise uses a assumed mean.

x	f	$x' = \frac{x - 1\ 125}{250}$	fx'	fx'^2
375	8.8	-3	-26.4	79.2
625	15.3	-2	-30.6	61.2
875	23.8	-1	-23.8	23.8
1 125	25.9	0	0	0
1 375	18.7	+1	+18.7	18.7
1 750	5.2	+2.5	+13.0	32.5
2 250	2.3	+4.5	+10.35	46.575
	100.0		-38.75	261.975

$A = 1125,\ C = 250$

$$\bar{x} = A = \frac{C\Sigma fx'}{\Sigma f} = 1\ 125 + 250 \times \frac{-38.75}{100}$$
$$= 1\ 125 - 96.875 = £1\ 028$$

$$\sigma = C\sqrt{\frac{\Sigma fx'^2}{\Sigma f} \quad \left(\frac{\Sigma fx'}{\Sigma f}\right)^2}$$
$$= 250\sqrt{\frac{261.975}{100} - \left(\frac{-38.75}{100}\right)^2}$$
$$= 250\sqrt{2.619\ 75 - 0.150\ 16}$$
$$= 250\sqrt{2.469\ 59} = 250 \times 1.571\ 5 = £393$$

3.15 a i Range = largest item − smallest item. It is easy to calculate and easy to understand. It is a suitable measure when the number of observations is small. It is particularly useful in quality control procedures. Not suitable when the sample size is large.

ii Quartile deviation $= \frac{\text{upper quartile} - \text{lower quartile}}{2}$

The quartile deviation measures the average deviation of the quartiles from the median. It is a relatively easy measure to find and understand. Useful when the distribution is skew (see Chapter 4). Disadvantage is that it is based on the middle 50% of the observations and thus ignores observations outside this range. It is not very amenable for further theoretical work.

Note: The range and the quartile deviation are based on only some of the observations; the mean deviation and standard deviation are based on all the observations.

iii Mean deviation is the average arithmetical difference of observations from the mean. It is easy to understand and to calculate. It is not used much and is not very amenable for further theoretical work.

iv See 3.12(a).(ii). The standard deviation is the most important measure of dispersion. It is very important in statistical inference (see Chapter 10). The standard deviation is however affected by extreme items, for example in income distributions. It is the most difficult masure of dispersion to calculate and it is not easy to comprehend.

b The observations put in order of size are 8, 9, 9, 12, 18, 18, 24.

Range = 24 – 8 = 16, Q_1 = 9, Q_3 = 18.

$$\text{Quartile deviation} = \frac{18-9}{2} = 4.5$$

$\Sigma x = 8 + 9 + 9 + 12 + 18 + 18 + 24 = 98.$

$$\bar{x} = \frac{\Sigma x}{n} = \frac{98}{7} = 14$$

$$\text{Mean deviation} = \frac{\Sigma|x-\bar{x}|}{n}$$

$\Sigma|x-\bar{x}| = 6 + 5 + 5 + 2 + 4 + 4 + 10 = 36.$

$$\text{Mean deviation} = \frac{36}{7} = 5.14$$

$\Sigma(x-x)^2 = 36 + 25 + 25 + 4 + 16 + 16 + 100 = 222$

$$\text{Standard deviation} = \sqrt{\frac{\Sigma(x-\bar{x})^2}{n}} = \sqrt{\frac{222}{7}}$$

$$= \sqrt{31.714} = 5.63$$

or $$= \sqrt{\frac{\Sigma(x-\bar{x})^2}{n-1}} = \sqrt{\frac{222}{6}} =$$

$$= \sqrt{37} = 6.08$$

3.16 i

x	f	fx	fx^2
12.5	1	12.5	156.25
17.5	4	70.0	1 225.00
22.5	28	630.0	14 175.00
27.5	42	1 155.0	31 762.50
32.5	33	1 072.5	34 856.25
37.5	18	675.0	25 312.50
42.5	13	552.5	23 481.25
47.5	9	427.5	20 306.25
52.5	2	105.0	5 512.50
	150	4 700.0	156 787.50

$$\bar{x} = \frac{\Sigma fx}{\Sigma f} = \frac{4\,700}{150} = £31.33$$

$$\sigma = \sqrt{\frac{\Sigma fx^2}{\Sigma f} - \left(\frac{\Sigma fx}{\Sigma f}\right)^2} = \sqrt{\frac{156\,787.5}{150} - \left(\frac{4\,700}{150}\right)^2}$$

$$= \sqrt{1\,045.25 - 981.776} = \sqrt{63.474}$$

$$= £7.97$$

ii **a** If all the women have an increase of £4.40 per week, this will not affect the spread about the mean, thus there will be no change in the standard deviation. As all the women have an increase of £4.40 the mean will increase by £4.40.

b If all the women have an increase of 10%, this means that those women with above-average earnings will get a larger increase than those with below-average earnings. Hence the deviations from the mean will change — they will in fact change by 10%; thus the standard deviation will increase by 10%: £7.97 + 10% of £7.97 = £7.97 + 0.80 = £8.77. Likewise the mean will increase by 10%: £31.33 + 10% of £31.33 = £31.33 + 3.13 = £34.46.

3.17 $\bar{x} = \frac{\Sigma x}{n}$ $\quad 16.4 = \frac{\Sigma x}{30}$ $\quad \Sigma x = 16.4 \times 30 = 492$. Replace 31, 21 by 13, 12, means a net reduction of Σx by 52 – 25 = 27. Thus corrected Σx is 492 – 27 = 465.

Thus correct mean is $\frac{465}{30} = 15.5$.

$$\sigma^2 = \frac{\Sigma(x - \bar{x})^2}{n} = \frac{\Sigma x^2 - \frac{(\Sigma x)^2}{n}}{n}$$

Thus $3.52^2 = \frac{\Sigma x^2 - \frac{(492)^2}{30}}{30}$

Thus $\Sigma x^2 = 30 \times 3.52^2 + \frac{492^2}{30} = 8\,440.512$.

Replace 31^2, 21^2 by 13^2, 12^2.

Thus corrected $\Sigma x^2 = 8\,440.512 - 31^2 - 21^2 + 13^2 + 12^2 = 7\,351.512$

$$\sigma^2 = \frac{\Sigma x^2 - \frac{(\Sigma x)^2}{n}}{n} = \frac{7\,351.512 - \frac{465^2}{30}}{30} = 4.800\,4$$

Thus $\sigma = 2.19$ (If a divisor of $n - 1$ is used, $\sigma = 2.13$)

Chapter 4

4.1

x	f	fx	fx^2	Cumulative frequency
22.5	21	472.5	110 631.5	21
30	45	1 350	40 500	66 ← M
40	21	840	33 600	87
50	10	500	25 000	93
60	3	180	10 800	100
	100	3 342.5	120 531.5	

$$\bar{x} = \frac{\Sigma fx}{\Sigma f} = \frac{3\ 342.5}{100} = 33.43 \text{ years}$$

$$\sigma = \sqrt{\frac{\Sigma fx^2}{\Sigma f} - \left(\frac{\Sigma fx}{\Sigma f}\right)^2}$$

$$= \sqrt{\frac{120\ 531.5}{100} - \left(\frac{3\ 342.5}{100}\right)^2}$$

$$= \sqrt{1\ 205.315 - 1\ 117.231}$$

$$= \sqrt{88.084} = 9.39 \text{ years}$$

$$\text{Median} = 25 + (35 - 25)\ \frac{50 - 21}{66 - 21}$$

$$= 25 + 10 \times \frac{29}{45} = 25 + 6.44 = 31.44 \text{ years}$$

$$\text{Measure of skewness}\ \frac{3(\bar{x} - \text{Median})}{\sigma}\quad \frac{3(33.43 - 31.44)}{9.39}$$

$$= \frac{3 \times 1.99}{9.39} = 0.64$$

4.2

	Cumulative frequency			
Under	*Males*		*Females*	
20	6.4		1.1	
25	17.4	← Q_1	9.7	
35	40.1	← M	29.9	← Q_1
45	59.8	← Q_3	55.6	← M
55	80.2		84.0	← Q_3
60	88.8		93.3	
65	96.7		98.1	
70	100.0		100.0	

$$\frac{n}{2} = \frac{100}{2} = 50 \qquad \frac{n}{4} = \frac{100}{4} = 25. \qquad \frac{3n}{4} = \frac{3 \times 100}{4} = 75$$

Males

$$Q_1 = 25 + (35 - 25)\left(\frac{25 - 17.4}{40.1 - 17.4}\right) = 28.35 \text{ years}$$

$$M = 35 + (45 - 35)\left(\frac{50 - 40.1}{59.8 - 40.1}\right) = 40.03 \text{ years}$$

$$Q_3 = 45 + (55 - 45)\left(\frac{75 - 59.8}{80.2 - 59.8}\right) = 52.45 \text{ years}$$

Females

$$Q_1 = 25 + (35 - 25)\left(\frac{25 - 9.7}{29.9 - 9.7}\right) = 32.57 \text{ years}$$

$$M = 35 + (45 - 35)\left(\frac{50 - 29.9}{55.6 - 29.9}\right) = 42.82 \text{ years}$$

$$Q_3 = 45 + (55 - 45)\left(\frac{75 - 55.6}{84 - 55.6}\right) = 51.83 \text{ years}$$

Skewness

$$\text{Males} = \frac{Q_3 + Q_1 - 2M}{Q_3 - Q_1} = \frac{28.35 + 52.45 - 2 \times 40.03}{52.45 - 28.35}$$

$$= \frac{0.74}{24.1} = 0.031$$

$$\text{Females} = \frac{Q_3 + Q_1 - 2M}{Q_3 - Q_1} = \frac{51.83 + 32.57 - 2 \times 42.82}{51.83 - 32.57}$$

$$= \frac{-1.24}{19.26} = -0.064$$

Quartile deviation

$$\text{Males} = \frac{52.45 - 28.35}{2} = 12.05 \text{ years}$$

$$\text{Females} = \frac{51.83 - 32.57}{2} = 9.63 \text{ years}$$

4.3 As numbers are large, use an assumed mean.

x	$x' = \frac{x - A}{C}$	f	fx'	fx'^2
125	−4	3.8	−15.2	60.8
375	−3	5.7	−17.1	53.3
625	−2	7.2	−14.4	28.8
875	−1	12.3	−12.3	12.3
1 125	0	21.0	0	0
1 375	+1	16.7	+16.7	16.7
1 750	+2.5	20.5	+51.0	127.5
2 250	+4.5	7.2	+32.4	145.8
2 750	+6.5	3.8	+24.7	160.55
3 250	+8.5	1.9	+16.15	137.275
		100.0	+81.95	742.625

$$\bar{x} = A + C \times \frac{\Sigma fx'}{\Sigma f} = 1\,125 + 250 \times \frac{81.95}{100} = £1\,330$$

$$\sigma = C\sqrt{\frac{\Sigma fx'^2}{\Sigma f} - \left(\frac{\Sigma fx'}{\Sigma f}\right)^2} = 250\sqrt{\frac{742.625}{100} - \left(\frac{81.95}{100}\right)^2}$$

$$= 250\sqrt{6.75467} = £649$$

$$\frac{\sigma \times 100}{\bar{x}} = \frac{649 \times 100}{1\,330} = 49$$

x	$x' = \frac{x-A}{C}$	f	fx′	fx′²
125	−3	8.8	−26.4	79.2
375	−2	15.3	−30.6	61.2
625	−1	23.8	−23.8	23.8
875	0	25.9	0	0
1 125	+1	18.7	+18.7	18.7
1 375	+2	5.2	+10.4	20.8
1 750	+3.5	1.5	+ 5.25	18.375
2 250	+5.5	0.4	+ 2.2	12.1
2.750	+7.5	0.4	+ 3.0	22.5
		100.0	−41.25	256.675

$$\bar{x} = 875 + 250 \times \frac{(-41.25)}{100} = £772$$

$$\sigma = 250\sqrt{\frac{256.675}{100} - \left(\frac{(-41.25)}{100}\right)^2}$$

$$= 250\sqrt{2.39659} = £387$$

$$\frac{\sigma \times 100}{\bar{x}} = \frac{387 \times 100}{772} = 50$$

4.4

x	f	fx	fx²	Cumulative frequency	
0	5	0	0	5	
0.25	9	2.25	0.562 5	14	
0.75	11	8.25	6.187 5	25	
1.25	18	22.5	28.125	43	⟵ M
1.75	23	40.25	70.437 5	66	
2.25	16	36.0	81.0	82	
2.75	7	19.25	52.937 5	89	
3.25	5	16.25	52.812 5	94	
3.75	3	11.25	42.187 5	97	
4.25	3	12.75	54.187 5	100	
	100	168.75	388.437 5		

$$\bar{x} = \frac{\Sigma fx}{\Sigma f} = \frac{168.75}{100} = 1.69 \text{ minutes}$$

$$\sigma = \sqrt{\frac{\Sigma fx^2}{\Sigma f} - \left(\frac{\Sigma fx}{\Sigma f}\right)^2} = \sqrt{\frac{388.437\ 5}{100} - \left(\frac{168.75}{100}\right)^2}$$

$$= \sqrt{3.884\ 375 - 2.847\ 65} = \sqrt{1.036\ 7}$$

$$= 1.02 \text{ minutes}$$

$$M = 1.5 + (2 - 1.5)\ \frac{50 - 43}{66 - 43} = 1.65 \text{ minutes}$$

$$\text{Skewness} = \frac{3(\bar{x} - M)}{\sigma}\quad \frac{3(1.69 - 1.65)}{1.02} = 0.12$$

4.5

x	f	fx	fx^2
17	1.1	18.7	317.9
19	1.1	20.9	397.1
22.5	8.6	193.5	4 353.75
27.5	14.9	409.75	11 268.125
32.5	18.2	591.5	19 223.75
37.5	18.1	678.75	25 453.125
42.5	14.2	603.5	25 648.75
47.5	9.7	460.75	21 885.625
55	9.4	517.0	28 435.0
70	4.0	280.0	19 600.0
90	0.7	63.0	5 670.0
	100.0	3 837.35	162 253.125

$$\bar{x} = \frac{\Sigma f x}{\Sigma f} = \frac{3\ 837.35}{100} = £38.37$$

$$\sigma = \sqrt{\frac{\Sigma f x^2}{\Sigma f} - \left(\frac{\Sigma f x}{\Sigma f}\right)^2} = \sqrt{\frac{162\ 253.125}{100} - \left(\frac{3\ 837.35}{100}\right)^2}$$

$$= \sqrt{1\ 622.531\ 25 - 1\ 472.525\ 5} = \sqrt{150.006} = £12.25$$

$$\text{Skewness} = \frac{3(\bar{x} - M)}{\sigma} = \frac{3(38.37 - 36.60)}{12.25} = 0.433$$

Distribution is positively skew. This is to be expected in income distributions.

The factors affecting the accuracy of the mean and standard deviation relate to the assumptions made regarding the central values. In particular we have made assumptions about the central values for the open-ended classes, but in this exercise any errors caused by this assumption are likely to be small since the frequencies of the open-ended classes are small. We have also assumed that the central value for each group is always at the midpoint of the range; this may not be the case. It is possible to deal with the this assumption by using Sheppard's correction for grouping. Sheppard's correction is explained in more advanced books on statistics.

4.6

Under	*Cumulative frequency*	
65	16	
70	41	← Q_1
75	65	← M
80	83	← Q_3
85	94	
90	99	
100	100	

$$\frac{n}{2} = \frac{100}{2} = 50 \qquad \frac{n}{4} = \frac{100}{4} = 25 \qquad \frac{3n}{4} = \frac{3 \times 100}{4} = 75$$

$$Q_1 = 65 + (70-65)\left(\frac{25-16}{41-16}\right) = 66.8 \text{ years}$$

$$Q_3 = 75 + (80-75)\left(\frac{75-65}{83-65}\right) = 77.78 \text{ years}$$

$$M = 70 + (75-70)\left(\frac{50-41}{65-41}\right) = 71.88 \text{ years}$$

		Men		*Women*
$\frac{Q_3-Q_1}{2}$	=	$\frac{76.4-68.3}{2}$	=	$\frac{77.78-66.8}{2}$
		= 4.05 years		= 5.5 years
$\frac{Q_3+Q_1-2M}{Q_3-Q_1}$	=	$\frac{76.4+68.3-2\times71.9}{76.4-68.3}$	=	$\frac{77.78+66.8-2\times71.88}{77.78-66.8}$
		= 0.111		= 0.075

4.7

Under	*Cumulative frequency*	
10	21	
15	78	← D_1
20	135	← Q_1
30	247	
40	380	← M
50	521	
60	634	← Q_3
80	784	
100	861	
150	936	

$$\frac{n}{10} = \frac{936}{10} = 93.6, \quad \frac{n}{2} = \frac{936}{2} = 468,$$

$$\frac{n}{4} = \frac{936}{4} = 234, \quad \frac{3n}{4} = \frac{3\times936}{4} = 702$$

$$D_1 = 15 + (20-15)\left(\frac{93.6-78}{135-78}\right) = £16.37$$

$$Q_1 = 20 + (30-20)\left(\frac{234-135}{247-135}\right) = £28.84$$

$$Q_3 = 60 + (80-60)\left(\frac{702-634}{784-634}\right) = £69.07$$

$$M = 40 + (50-40)\left(\frac{468-380}{521-380}\right) = £46.24$$

		South West Region		*Greater London Area*
$\frac{Q_3-Q_1}{2}$	=	$\frac{69.07-28.84}{2}$	=	$\frac{76.30-28.00}{2}$
		= £20.11		= £24.15
$\frac{Q_3+Q_1-2M}{Q_3-Q_1}$	=	$\frac{69.07+28.84-2\times46.24}{69.07-28.84}$	=	$\frac{76.30+28.00-2\times52.50}{76.30-28.00}$
		= 0.135		= -0.014

For part (c) see answer to 2.10.

4.8

1961				*1971*			
x	f	fx	fx^2	x	f	fx	fx^2
1	1	1	1	1	2	2	2
2	5	10	20	2	4	8	16
3	12	36	108	3	9	27	81
4	27	108	432	4	23	92	368
5	35	175	875	5	30	150	750
6	13	78	468	6	23	138	828
7	3	21	147	7	5	35	245
9	4	36	324	9	4	36	324
	100	465	2 375		100	488	2 614

1961

$$\bar{x} = \frac{\Sigma fx}{\Sigma f} = \frac{465}{100} = 4.65 \text{ rooms}$$

$$\sigma = \sqrt{\frac{\Sigma fx^2}{\Sigma f} - \left(\frac{\Sigma fx}{\Sigma f}\right)^2}$$

$$= \sqrt{\frac{2\,375}{100} - \left(\frac{465}{100}\right)^2}$$

$$= \sqrt{2.1275}$$

$$= 1.46 \text{ rooms}$$

$$\frac{100\sigma}{\bar{x}} = \frac{1.46 \times 100}{4.65}$$

$$= 31.4$$

1971

$$\frac{488}{100} = 4.88 \text{ rooms}$$

$$= \sqrt{\frac{2\,614}{100} - \left(\frac{488}{100}\right)^2}$$

$$= \sqrt{2.3256}$$

$$= 1.52 \text{ rooms}$$

$$= \frac{1.52 \times 100}{4.88}$$

$$= 31.2$$

4.9 **a** First group central value 1 800. This assumes that the class interval of 400 which follows this group is applicable to the under 2 000 group. In view of the fact that some television tubes will probably have a very short life, this assumption is very subjective. Last group 6 500. This assumes that the class intervals of the previous two groups are applicable to the final group.

b With large numbers we need to deal with 'overflow' problems. The solution uses an assumed mean. If you work in hundreds of hours, then you should obtain $\Sigma fx = 33\ 574$, $\Sigma fx^2 = 1\ 216\ 744$ and the same result as below when you multiply your values of $\bar{x}$ and σ by 100.

x	$x' = \frac{x - 3000}{400}$	f	fx'	fx'^2
1 800	−3	35	−105	315
2 200	−2	115	−230	460
2 600	−1	142	−142	142
3 000	0	193	0	0
3 400	+1	174	+174	174
3 800	+2	157	+314	628
4 500	+3.75	120	+450	1 687.5
5 500	+6.25	51	+318.75	1 992.187 5
6 500	+8.75	13	+113.75	995.312 5
		1 000	+893.5	6 394.0

$A = 3\,000 \quad C = 400$

$$\bar{x} = A + C\frac{\Sigma fx'}{\Sigma f} = 3\,000 + 400 \times \frac{893.5}{1\,000}$$

$$= 3\,000 + 357.4 = 3\,357 \text{ hours}$$

$$\sigma = C\sqrt{\frac{\Sigma fx'^2}{\Sigma f} - \left(\frac{\Sigma fx'}{\Sigma f}\right)^2}$$

$$= 400\sqrt{\frac{6\,394}{1\,000} - \left(\frac{893.5}{1\,000}\right)^2} = 946 \text{ hours}$$

$$\text{Coefficient of variation} = \frac{\sigma 100}{\bar{x}}$$

$$= \frac{946 \times 100}{3\,357} = 28.18$$

c The standard deviation is a measure of dispersion. In the case of the normal distribution (see Chapter 9), 95% of the observations lie within 1.96 standard deviations of the mean. In statistical inference (see Chapter 10) the standard deviation plays an important part in sampling distributions.

The coefficient of variation measures relative variation. It is the ratio of the standard deviation to the mean. It is useful for comparing the relative variation of different distributions and is particularly useful when the units differ, for example heights and weights, or to measure the relative variation of incomes in different countries where the units of currency differ.

4.10 Arrange in order: 2, 4, 6, 8, 9, 9, 11, 15, 16, 20. The middle item is 9. Thus the median is 9. $\Sigma x = 2 + 4 + 6 + 8 + 9 + 9 + 11 + 15 + 16 + 20 = 100$

$$\text{Mean} = \bar{x} = \frac{\Sigma x}{n} = \frac{100}{10} = 10$$

$$\text{Mean deviation} = \frac{\Sigma|x - \bar{x}|}{n}.$$

$$\text{Standard deviation} = \sqrt{\frac{\Sigma(x - \bar{x})^2}{n}}.$$

$$\Sigma|x - \bar{x}| = 8 + 6 + 4 + 2 + 1 + 1 + 1 + 5 + 6 + 10 = 44$$

$$\text{Mean deviation} = \frac{44}{10} = 4.4$$

$\Sigma(x - \bar{x})^2 = 8^2 + 6^2 + 4^2 + 2^2 + 1^2 + 1^2 + 1^2 + 5^2 + 7^2 + 10^2 = 284$

$$\text{Standard deviation} = \sqrt{\frac{284}{10}} = 5.33$$

$$\text{Measure of skewness} = \frac{3(\bar{x} - M)}{\sigma} = \frac{3(10 - 9)}{5.33} = 0.563$$

4.11 Range $= 18 - 8 = 10$

$\Sigma x = 8 + 9 + 17 + 11 + 18 + 16 + 14 + 13 + 12 + 9 = 127$

$$\bar{x} = \frac{\Sigma x}{n} = \frac{127}{10} = 12.7$$

$$\Sigma(x - \bar{x})^2 = 4.7^2 + 3.7^2 + 4.3^2 + 1.7^2 + 5.3^2 + 3.3^2 + 1.3^2 + 0.3^2 + 0.7^2 + 3.7^2 = 112.1$$

$$\text{Standard deviation} = \sqrt{\frac{\Sigma(x - \bar{x})^2}{n}} = \sqrt{\frac{112.1}{10}} = 3.35$$

$$\text{Coefficient of variation} = \frac{\sigma 100}{\bar{x}} = \frac{3.35 \times 100}{12.7} = 26.4$$

Chapter 5

5.1

X	Y	X^2	XY
5	1.7	25	8.5
7	2.4	49	16.8
9	2.8	81	25.2
11	3.4	121	37.4
13	3.7	169	48.1
15	4.4	225	66.0
60	18.4	670	202.0

$$\bar{X} = \frac{\Sigma X}{n} = \frac{60}{6} = 10 \qquad \bar{Y} = \frac{\Sigma Y}{n} = \frac{18.4}{6} = 3.067$$

$$b = \frac{n\Sigma YX - \Sigma Y \Sigma X}{n\Sigma X^2 - (\Sigma X)^2} = \frac{6 \times 202 - 18.4 \times 60}{6 \times 670 - (60)^2} = \frac{108}{420}$$

$$= 0.257\,1$$

$$a = \bar{Y} - b\bar{X} = 3.067 - 0.2571 \times 10 = 0.495$$

$$Y = a + bX \qquad Y = 0.495 + 0.257X$$

5.2

X	Y	X^2	XY
1	6.2	1	6.2
2	6.9	4	13.8
3	7.1	9	21.3
4	11.1	16	44.4
5	14.4	25	72.0
6	14.9	36	89.4
7	14.9	49	104.3
28	75.5	140	351.4

$$\bar{X} = \frac{\Sigma X}{n} = \frac{28}{7} = 4 \qquad \bar{Y} = \frac{\Sigma Y}{n} = \frac{75.5}{7} = 10.786$$

$$b = \frac{n\Sigma YX - \Sigma Y\Sigma X}{n\Sigma X^2 - (\Sigma X)^2} = \frac{7 \times 351.4 - 75.5 \times 28}{7 \times 140 - (28)^2} = \frac{345.8}{196}$$
$$= 1.764\,3$$
$$a = \overline{Y} - b\overline{X} = 10.786 - 1.7643 \times 4 = 3.729$$
$$Y = a + bX \qquad Y = 3.73 + 1.76X$$

In 1977 $X = 9$, $Y = 3.73 + 1.76 \times 9 = 19.6$

Your scatter diagram should show that the data do not lie on a straight line. There was a big increase in house prices between 1971 and 1973, but between 1973 and 1975 hardly any increase. Thus any forecast based on the projection of a straight line which is not a good fit to the data is unlikely to be very realistic. There are a number of factors which affect house prices; availability and cost of mortgages are probably the most important factors. Thus a forecast of the movement of house prices would involve also a forecast of the future availability of mortgages, and it is unlikely that mortgage availability can be forecast very accurately.

5.3

X	Y	X^2	Y^2	XY
6	2.5	36	6.25	15
5	2.4	25	5.76	12
5	3.3	25	10.89	16.5
8	3.3	64	10.89	26.4
12	3.5	144	12.25	42
17	3.7	289	13.69	62.9
19	3.9	361	15.21	74.1
18	3.6	324	12.96	64.8
14	3.4	196	11.56	47.6
11	3.1	121	9.61	34.1
115	32.7	1585	109.07	395.4

$$r = \frac{n\Sigma XY - \Sigma Y\Sigma X}{\sqrt{(n\Sigma X^2 - (\Sigma X)^2)(n\Sigma Y^2 - (\Sigma Y)^2)}}$$
$$= \frac{10 \times 395.4 - 32.7 \times 115}{\sqrt{(10 \times 1\,585 - (115)^2)(10 \times 109.07 - (32.7)^2)}}$$
$$= \frac{193.5}{\sqrt{2\,625 \times 21.41}} = \frac{193.5}{237.1} = 0.816$$

5.4

	A	B	C	D	E	F	G	H	I	J	
	7	3	5	8	2	10	1	9	6	3	$\Sigma d = 0$
	8	4	6	5	1	10	3	9	6	2	(checks
d	-1	-1	-1	+3	+1	0	-2	0	0	+1	arithmetic)
Σd^2	1 +	1 +	1 +	9 +	1 +	0 +	4 +	0 +	0 +	1	= 18

$$r = 1 - \frac{6\Sigma d^2}{n^3 - n} = 1 - \frac{6 \times 18}{10^3 - 10} = 1 - 0.109 = 0.891$$

5.5 **a**

X	Y	X^2	Y^2	XY
5	12	25	144	60
7	9	49	81	63
8	14	64	196	112
11	20	121	400	220
13	13	169	169	169
15	14	225	196	210
16	22	256	484	352
19	21	361	441	399
22	29	484	841	638
23	28	529	784	644
139	182	2 283	3 736	2 867

$$r = \frac{n\Sigma XY - \Sigma Y\Sigma X}{\sqrt{(n\Sigma X^2 - (\Sigma X)^2)(n\Sigma Y^2 - (\Sigma Y)^2)}}$$

$$= \frac{10 \times 2\,867 - 182 \times 139}{\sqrt{(10 \times 2\,283 - (139)^2)(10 \times 3\,736 - (182)^2)}}$$

$$= \frac{3\,372}{\sqrt{3\,509 \times 4\,236}} = \frac{3\,372}{3\,855.9} = 0.875$$

b

Rank X	Rank Y	d	d^2
1	2	-1	1
2	1	+1	1
3	$4\frac{1}{2}$	$-1\frac{1}{2}$	$2\frac{1}{4}$
4	6	-2	4
5	3	+2	4
6	$4\frac{1}{2}$	$+1\frac{1}{2}$	$2\frac{1}{4}$
7	8	-1	1
8	7	+1	1
9	10	-1	1
10	9	+1	1
		0	$18\frac{1}{2}$

$$r = 1 - \frac{6\Sigma d^2}{n^3 - n}$$

$$= 1 - \frac{6 \times 18.5}{10^3 - 10}$$

$$= 1 - 0.112$$

$$= 0.888$$

Both rank and product moment correlation coefficients show a high positive value. When data are placed into ranks some information is lost; thus when we have a choice, as in this exercise, the product moment correlation coefficient is to be preferred as it utilises all the data. One advantage of using a rank correlation coefficient is of course it is quicker to calculate.

5.6

X	Y	X^2	Y^2	XY
2	9	4	81	18
4	7	16	49	28
6	4	36	16	24
8	5	64	25	40
10	3	100	9	30
12	2	144	4	24
42	30	364	184	164

$$\overline{X} = \frac{\Sigma X}{n} = \frac{42}{6} = 7 \qquad \overline{Y} = \frac{\Sigma Y}{n} = \frac{30}{6} = 5$$

$$b = \frac{n\Sigma YX - \Sigma Y\Sigma X}{n\Sigma X^2 - (\Sigma X)^2} = \frac{6 \times 164 - 30 \times 42}{6 \times 364 - (42)^2} = \frac{-276}{420}$$

$$= \underline{-0.657}$$

$$a = \overline{Y} - b\overline{X} = 5 - (-0.657) \times 7 = 5 + 4.60 = 9.60$$

$$Y = a + bX, \quad Y = 9.60 - 0.657X$$

5.7 Using the table of Exercise 5.6,

$$r = \frac{n\Sigma YX - \Sigma Y\Sigma X}{\sqrt{(n\Sigma X^2 - (\Sigma X)^2)(n\Sigma Y^2 - (\Sigma Y)^2)}}$$

$$= \frac{6 \times 164 - 30 \times 42}{\sqrt{(6 \times 364 - (42)^2)(6 \times 184 - (30)^2)}}$$

$$= \frac{-276}{\sqrt{420 \times 204}} = \frac{-276}{292.7} = -0.943$$

5.8 Let March 1973 = 1, April 1973 = 2 and so on

X	Y	X^2	XY
1	2.17	1	2.17
2	2.21	4	4.42
3	2.23	9	6.69
4	2.26	16	9.04
5	2.28	25	11.40
6	2.33	36	13.98
7	2.36	49	16.52
8	2.40	64	19.20
9	2.45	81	22.05
10	2.48	100	24.80
11	2.43	121	26.73
66	25.60	506	157.00

$$\overline{X} = \frac{\Sigma X}{n} = \frac{66}{11} = 6 \qquad \overline{Y} = \frac{\Sigma Y}{n} = \frac{25.60}{11} = 2.327$$

$$b = \frac{n\Sigma YX - \Sigma Y\Sigma X}{n\Sigma X^2 - (\Sigma X)^2} = \frac{11 \times 157 - 25.6 \times 66}{11 \times 506 - (66)^2} = \frac{37.4}{1210}$$

$$= 0.030\,91$$

$a = \overline{Y} - b\overline{X} = 2.327 - 0.030\,91 \times 6 = 2.142$
$Y = a + bX, \quad Y = 2.142 + 0.030\,9X$
In February 1974 X = 12, Y = 2.142 + 0.030 9 × 12 = 2.51
The data are given seasonally unadjusted so it could be that in February because of seasonal factors the higher purchase debt outstanding is always below the trend; this factor could account for the over-estimate. A scatter diagram and a plot of the regression line reveals that November and December 1973 were above the regression line and January 1974 was below the regression line. H.P. debt is also affected by Government fiscal policies. February 1974 was during the 3 day week at the time of the miners' strike of 1974 and this may have been another factor.

5.9 **(a)**

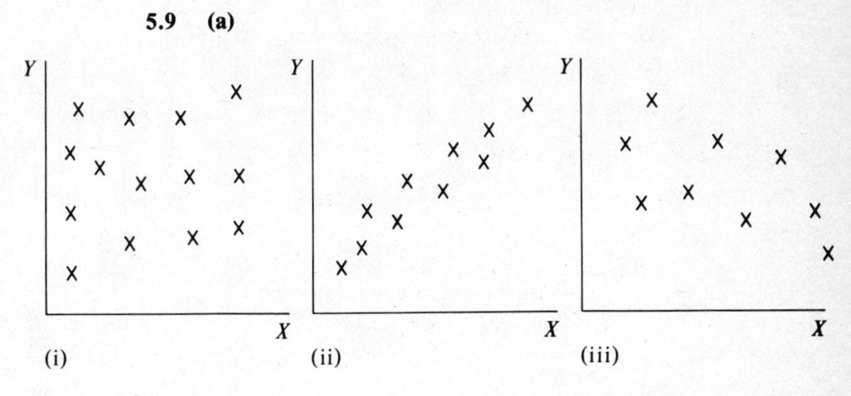

Figure A5.9

(b)

	A	B	C	D	E	F	G	H	I	J	
	3	7	10	2	9	1	5	4	6	8	
	1	10	9	4	8	3	2	7	5	6	
d	+2	−3	+1	−2	+1	−2	+3	−3	+1	+2	
Σd^2	4+	9+	1+	4+	1+	4+	9+	9+	1+	4	=46

$$r = 1 - \frac{6\Sigma d^2}{n^3 - n}$$
$$= 1 - \frac{6 \times 46}{10^3 - 10}$$
$$= 1 - 0.279$$
$$= 0.721$$

5.10

X	Y	X^2	XY
18	54	324	972
26	64	676	1 664
28	54	784	1 512
34	62	1 156	2 108
36	68	1 296	2 448
42	70	1 764	2 940

X	Y	X^2	XY
48	76	2 304	3 648
52	66	2 704	3 432
54	76	2 916	4 104
60	74	3 600	4 400
398	664	17 524	27 268

$$\overline{X} = \frac{\Sigma X}{n} = \frac{398}{10} = 39.8 \qquad \overline{Y} = \frac{\Sigma Y}{n} = \frac{664}{10} = 66.4$$

$$b = \frac{n\Sigma YX - \Sigma Y\Sigma X}{n\Sigma X^2 - (\Sigma X)^2} = \frac{10 \times 27\,268 - 664 \times 398}{10 \times 17\,524 - (398)^2}$$

$$= \frac{8\,408}{16\,836} = 0.499\,4$$

$$a = \overline{Y} - b\overline{X} = 66.4 - 0.4994 \times 39.8 = 46.5$$

$$Y = a + bX \qquad Y = 46.5 + 0.499X$$

When X = 70, Y = 46.5 + 0.499 × 70 = 81 = £81 000

5.11

X	Y	X^2	Y^2	XY	rank X	rank Y	d	d^2
0.4	1.8	0.16	3.24	0.72	$9\frac{1}{2}$	9	$+\frac{1}{2}$	$\frac{1}{4}$
0.4	1.6	0.16	2.56	0.64	$9\frac{1}{2}$	10	$-\frac{1}{2}$	$\frac{1}{4}$
0.6	2.2	0.36	4.84	1.32	$7\frac{1}{2}$	$7\frac{1}{2}$	0	0
1.0	3.0	1.00	9.00	3.00	$3\frac{1}{2}$	5	$-1\frac{1}{2}$	$2\frac{1}{4}$
0.7	2.2	0.49	4.84	1.54	6	$7\frac{1}{2}$	$-1\frac{1}{2}$	$2\frac{1}{4}$
0.6	2.4	0.36	5.76	1.44	$7\frac{1}{2}$	6	$+1\frac{1}{2}$	$2\frac{1}{4}$
0.8	3.2	0.64	10.24	2.56	5	4	+1	1
1.0	4.0	1.00	16.00	4.00	$3\frac{1}{2}$	$2\frac{1}{2}$	+1	1
1.7	5.3	2.89	28.09	9.01	1	1	0	0
1.6	4.0	2.56	16.00	6.40	2	$2\frac{1}{2}$	$-\frac{1}{2}$	$\frac{1}{4}$
8.8	29.7	9.62	100.57	30.63			0	$9\frac{1}{2}$

$$r = 1 - \frac{6\Sigma d^2}{n^3 - n}$$

$$= 1 - \frac{6 \times 9\frac{1}{2}}{10^3 - 10}$$

$$= 1 - 0.058$$

$$= 0.942$$

$$r = \frac{n\Sigma XY - \Sigma Y\Sigma X}{\sqrt{(n\Sigma X^2 - (\Sigma X)^2)(n\Sigma Y^2 - (\Sigma Y)^2)}}$$

$$= \frac{10 \times 30.63 - 29.7 \times 8.8}{\sqrt{(10 \times 9.62 - (8.8)^2)(10 \times 100.57 - (29.7)^2)}}$$

$$= \frac{44.94}{\sqrt{18.76 \times 123.61}} = \frac{44.94}{48.15} = 0.933$$

5.12 a

X	Y	X^2	XY
18	15.5	324	279
20	23.2	400	464
22	34.0	484	748
27	44.9	729	1 212.3
35	53.1	1 225	1 858.5
45	55.0	2 025	2 475
55	51.2	3 025	2 816
222	276.9	8 212	9 852.8

$$\overline{X} = \frac{\Sigma X}{n} = \frac{222}{7} = 31.714, \qquad \overline{Y} = \frac{\Sigma Y}{n} = \frac{276.9}{7} = 39.56$$

$$b = \frac{n\Sigma YX - \Sigma Y\Sigma X}{n\Sigma X^2 - (\Sigma X)^2} = \frac{7 \times 9\,852.8 - 276.9 \times 222}{7 \times 8\,212 - (222)^2}$$

$$= \frac{7\,497.8}{8\,200} = 0.914\,4$$

$$a = \overline{Y} - b\overline{X} = 39.56 - 0.914\,4 \times 31.714 = 10.56$$

$$Y = a + bX, \quad Y = 10.56 + 0.914X$$

b If all the non-manual workers have an increase of £6 per week, this will not affect the scatter since all the Y values increase by 6. The regression line will therefore move up parallel from its previous position; b will therefore remain the same, i.e. 0.914. Now as all the Y values increase by 6, $\overline{Y}$ will increase by 6. $a = \overline{Y} - b\overline{X}$. b and $\overline{X}$ remain unchanged, $\overline{Y}$ increases by 6, hence a increases by 6, and becomes 16.56.

If all the non-manual workers have an increase of 6%, each Y will increase by 6%. Thus $\overline{Y}$ will increase by 6%. Now $b = \frac{n\Sigma XY - \Sigma Y\Sigma X}{n\Sigma X^2 - (\Sigma X)^2}$ As the age of non-manual workers will not change in the context of the question, X's will remain constant, thus the denominator of the expression for b will not change but the numerator will change, since each Y increases by 6%; thus b will increase by 6%. The new value of b = 0.9144 + 6% of 0.9144 = 0.969.

5.13

X	Y	X^2	Y^2	XY
37	113	1 369	12 769	4 181
33	68	1 089	4 624	2 244
40	193	1 600	37 249	7 720
37	124	1 369	15 376	4 588
26	64	676	4 096	1 664
26	74	676	5 476	1 924
39	93	1 521	8 649	3 627
238	729	8 300	88 239	25 948

ii
$$r = \frac{n\Sigma XY - \Sigma Y\Sigma X}{\sqrt{(n\Sigma X^2 - (\Sigma X)^2)(n\Sigma Y^2 - (\Sigma Y)^2)}}$$
$$= \frac{7 \times 25\,948 - 729 \times 238}{\sqrt{(7 \times 8\,300 - (238)^2)(7 \times 88\,239 - (729)^2)}}$$
$$= \frac{8\,134}{\sqrt{1\,456 \times 86\,232}} = \frac{8\,134}{11\,205} = 0.726$$

iii The value of the correlation coefficient of 0.726 shows that there is some relationship between the two variables. It does not, of course, show that there is a causative relationship between divorce rates and females as percentage of total economically active. The value of the coefficient of determination is $r^2 = 0.726^2 = 0.527$. This suggests that about 53% of the variation in divorce rates can be accounted for by the percentage of females of the total of those who are economically active. Assuming that the relationship is a causative one, this indicates that there are other factors to account for the remaining 47%.

iv See 5.14(c) for a discussion of correlation analysis.

In regression analysis we are attempting to find a relationship between variables. In elementary work we are trying to find the equation of the straight line which best fits the data. The use of the regression line for forecasting purposes presents problems, because this assumes that the line when extrapolated will represent future values. It is possible to find the equation of a straight line which is a good fit to the data, i.e. a high value of r, yet the data lies on a curve (data should always be plotted on a scatter diagram to reduce this risk). Clearly in this situation a straight line is unlikely to give accurate forecasts. It is possible in regression analysis to find the equation of a curve that best fits the data. It is also possible to find the regression equation when there are a number of explanatory variables — this is called multiple regression analysis. A discussion of these topics is outside the scope of this book.

5.14 a

X	Y	X^2	Y^2	XY
39	42	1 521	1 764	1 638
42	44	1 764	1 936	1 848
49	50	2 401	2 500	2 450
53	61	2 809	3 721	3 233
61	79	3 721	6 241	4 819
57	70	3 249	4 900	3 990
68	98	4 624	9 604	6 664
135	133	18 225	17 689	17 955
504	577	38 314	48 355	42 597

$$r = \frac{n\Sigma XY - \Sigma Y\Sigma X}{\sqrt{(n\Sigma X^2 - (\Sigma X)^2)(n\Sigma Y^2 - (\Sigma Y)^2)}}$$

$$= \frac{8 \times 42\,597 - 504 \times 577}{\sqrt{(8 \times 38\,314 - 504^2)(8 \times 48\,355 - 577^2)}}$$

$$= \frac{49\,968}{\sqrt{52\,496 \times 53\,911}} = \frac{49\,968}{53\,199} = 0.939$$

b This value of *r* is high, but is it likely to be a causative relationship? In this case both tax relief and housing subsidy are likely to have been affected by inflation. This is an example when both variables are related to a third variable — inflation.

c The correlation coefficient is used to measure the degree of association between two variables. A high positive value of *r* indicates that if one variable increases the other variable increases; a high negative value of *r* indicates that if one variable increases the other variable decreases. The correlation coefficient does not, however, prove a causative relationship. In some cases both variables increase with time but are not in any way related; part (a) of this question is probably a typical example of this situation. In this case both have increased with time because of inflation, and as both variables relate to housing there is the possibility of thinking that because $r = 0.939$, a high value, housing subsidy and tax relief to mortgagors have a causative relationship.

The correlation coefficient can also be used to show how good a fit a straight line — a regression line — is to the data. A low value of *r* would indicate a poor fit and thus the regression line would have limited use in forecasting.

The coefficient of determination $= r^2$ shows how much of the variation of the dependent variable Y can be accounted for by the explanatory variable X. Thus if $r = 0.9$, $r^2 = 0.81$, hence 81% of the variation of Y can be 'explained' by X and thus 19% of variation of Y is due to other factors. With a large sample, say $n = 100$, a small value of $r = 0.3$ is significant at 1%, but as $r^2 = 0.09$, this means that only 9% of the variation of Y is 'explained' by X. Clearly small values of *r*, even though significant, need to be interpreted with caution. The coefficient of determination is helpful in that it shows how much of the variation of Y is explained by X and thus the investigator can search for other explanatory variables to 'explain' the remaining variation of Y.

Chapter 6

6.1 $\Sigma p_n = 54 + 2 + 16\frac{1}{2} + 10 = 82\frac{1}{2}$

$\Sigma p_o = 43\frac{1}{2} + 1\frac{1}{2} + 11 + 7 = 63$

$$\frac{\Sigma p_n}{\Sigma p_o} \times 100 = \frac{82\frac{1}{2}}{63} \times 100 = 130.95$$

6.2

p_o	q_o	p_n	q_n	p_oq_o	p_nq_o	p_oq_n	p_nq_n
35	3	80	2	105	240	70	160
2	2	4.5	3	4	9	6	13.5
5	$1\frac{1}{2}$	9	1	7.5	13.5	5	9
				116.5	262.5	81	182.5

$$\frac{\Sigma p_nq_o}{\Sigma p_oq_o} \times 100 = \frac{262.5}{116.5} \times 100 = 225.3$$

$$\frac{\Sigma p_nq_n}{\Sigma p_oq_n} \times 100 = \frac{182.5}{81} \times 100 = 225.3$$

6.3

$$\frac{\Sigma q_np_o}{\Sigma q_op_o} \times 100 = \frac{81}{116.5} \times 100 = 69.5$$

$$\frac{\Sigma q_np_n}{\Sigma q_op_n} \times 100 = \frac{182.5}{262.5} \times 100 = 69.5$$

6.4

p_o	p_oq_o	q_o	p_n	p_nq_n	q_n	p_oq_n	p_nq_o
8	36	4.5	9.9	148.5	15	120	44.55
16.6	49.8	3	29.3	58.6	2	33.2	87.9
12.5	50	4	17	51	3	37.5	68
	135.8			258.1		190.7	200.45

$$\frac{\Sigma p_nq_o}{\Sigma p_oq_o} \times 100 = \frac{200.45}{135.8} \times 100 = 147.6$$

$$\frac{\Sigma p_nq_n}{\Sigma p_oq_n} \times 100 = \frac{258.1}{190.7} \times 100 = 135.3$$

6.5

W	I	$W \times I$
56	76	4 256
84	129	10 836
7	149	1 043
64	160	10 240
313	129	40 377
60	114	6 840
87	121	10 527
29	130	3 770
21	130	2 730

W	I	$W \times I$
59	122	7 198
25	159	3 975
127	115	14 605
68	116	7 888
1 000		124 285

a $\dfrac{\Sigma(W \times I)}{\Sigma W} = \dfrac{124\ 285}{1\ 000} = 124.3$

b For manufacturing industries

$$\Sigma W = 1\ 000 - (56 + 127 + 68)$$
$$= 1\ 000 - 251 = 749$$
$$\Sigma(W \times I) = 124\ 285 - (4\ 256 + 14\ 605 + 7\ 888)$$
$$= 124\ 285 - 26\ 749$$
$$= 97\ 536$$
$$\frac{\Sigma(W \times I)}{\Sigma W} = \frac{97\ 536}{749} = 130.2$$

6.6

W	I	$W \times I$
253	106.1	26 843.3
70	110.7	7 749.0
43	120.3	5 172.9
124	105.1	13 032.4
52	115.7	6 016.4
64	109.5	7 008.0
91	110.9	10 091.9
135	112.7	15 214.5
63	113.3	7 137.9
54	109.3	5 902.2
51	110.4	5 630.4
1 000		109 798.9

$$\frac{\Sigma(W \times I)}{\Sigma W} = \frac{109\ 798.9}{1\ 000} = 109.8$$

$$\frac{Y}{191.8} = \frac{109.8}{100} \qquad Y = \frac{191.8 \times 109.8}{100} = 210.6$$

6.7

p_o	q_o	p_n	q_n	p_oq_o	p_nq_o	p_nq_n	p_oq_n
30	350	35	300	10 500	12 250	10 500	9 000
22	300	23	350	6 600	6 900	8 050	7 700
15	80	18	80	1 200	1 440	1 440	1 200
10	60	11	80	600	660	880	800
				18 900	21 250	20 870	18 700

i $\dfrac{\Sigma p_nq_o}{\Sigma p_oq_o} = 100 = \dfrac{21\ 250}{18\ 900} \times 100 = 112.4$

$$\frac{\Sigma p_n q_n}{\Sigma p_o q_n} \times 100 = \frac{20\,870}{18\,700} \times 100 = 111.6$$

ii A base-weighted index usually understates the increase and a current-weighted index usually overstates the increase. In this example male wages have increased at a higher proportionate rate than female wages. The company has increased the employment of female employees and decreased the employment of male employees, i.e. the company has followed the procedure we would expect from economic analysis (see also Worked Example No. 6.2).

6.8

W	I	$W \times I$
37	87	3 219
84	103	8 652
65	115	7 475
57	78	4 446
319	99	31 581
76	103	7 828
144	103	14 832
146	91	13 286
72	112	8 064
1 000		99 383

i $\frac{\Sigma(W \times I)}{\Sigma W} = \frac{99\,383}{1\,000} = 99.4$

ii $\Sigma W = 1\,000 - (37 + 146 + 72) = 1\,000 - 255 = 745$

$\Sigma(W \times I) = 99\,383 - (3\,219 + 13\,286 + 8\,064)$

$= 99\,383 - 24\,569 = 74\,814$

$\frac{\Sigma(W \times I)}{\Sigma W} = \frac{74\,814}{745} = 100.4$

iii $\frac{Y}{124.2} = \frac{99.4}{100} \quad Y = \frac{124.2 \times 99.4}{100} = 123.5$

6.9

q_o	$p_o q_o$	q_n	$p_n q_n$	p_n	$p_n q_o$
318	14.6	256	14.0	0.054 69	17.4
210	8.5	193	9.0	0.046 63	9.8
167	24.2	158	25.2	0.159 5	26.6
113	13.8	145	17.8	0.122 8	13.9
	61.1				67.7

$$\frac{\Sigma p_n q_o}{\Sigma p_o q_o} \times 100 = \frac{67.7}{61.1} \times 100 = 111$$

6.10

p_o	q_o	p_n	q_n	p_oq_o	p_nq_n	p_oq_n	p_nq_o
1.8	3	2.2	4	5.4	8.8	7.2	6.6
4.5	1	3.9	1	4.5	3.9	4.5	3.9
28.9	1	38.7	$\frac{1}{2}$	28.9	19.35	14.45	38.7
48.4	2	58.7	3	96.8	176.1	145.2	117.4
21.6	1	28.3	$1\frac{1}{2}$	21.6	42.45	32.4	28.3
3.5	5	4.3	6	17.5	25.8	21.0	21.5
				174.7	276.40	224.75	216.4

i $\frac{\Sigma p_nq_o}{\Sigma p_oq_o} \times 100 = \frac{216.4}{174.7} \times 100 = 123.9$

ii $\frac{\Sigma p_nq_n}{\Sigma p_oq_n} \times 100 = \frac{276.40}{224.75} \times 100 = 123.0$

a See Worked Example 6.2.

6.11 a There are two main components in index number construction: (i) a set of weights; (ii) price relatives. In practice the weights are proportional to the expenditure at some period in the past. The expenditure pattern changes as a result of two factors: (i) the greater wealth of the community — a wealthier community tends to spend less on food and thus more on other commodities; (ii) quality changes and the introduction of new technology — colour television sets were not purchased before 1967 but now the majority of households have a colour television set. The spending pattern today is therefore very different from 20 years ago; thus comparing retail prices today with 20 years ago is not comparing like with like. It is well known that a Laspeyres-type index with weights fixed 20 years ago overstates the increase in prices. The current general index of retail prices changes weights each January and for the remainder of the year prices are compared with those of January. The index is then chained back to the base date. With this procedure we are clearly not comparing like with like, but we are avoiding the overstated increase that usually accompanies a base-weighted index. A fixed-weighted index is cheaper since it is only necessary to calculate the weights once. A chain base index is much more expensive, since weights have to be calculated each year.

b

W	I	$W \times I$
228	152.1	34 678.8
81	150.9	12 222.9
46	162.8	7 488.8
112	135.8	15 209.6
56	169.4	9 486.4
75	141.2	10 590.0
84	134.9	11 331.6

W	I	$W \times I$
140	156.9	21 966.0
74	154.2	11 410.8
57	154.9	8 829.3
47	148.3	6 970.1
1 000		150 184.3

$$\frac{\Sigma(W \times I)}{\Sigma W} = \frac{150\ 184.3}{1\ 000} = 150.2$$

Let Y be the value of the savings certificate in February 1976:

$$\frac{Y}{500} = \frac{150.2}{138.5} \qquad Y = \frac{500 \times 150.2}{138.5} = £542.24$$

6.12

Food			Housing		
$p_n q_n$	$p_o q_n$	$\frac{p_n q_n}{p_o q_n} \times 100$ = index	$p_n q_n$	$p_o q_n$	$\frac{p_n q_n}{p_o q_n} \times 100$ = index
6 365	6 365	$\frac{6\ 365}{6\ 365} \times 100 = 100$	4 181	4 181	$\frac{4\ 181}{4\ 181} \times 100 = 100$
6 964	6 362	$\frac{6\ 964}{6\ 362} \times 100 = 109.5$	4 719	4 287	$\frac{4\ 719}{4\ 287} \times 100 = 110.1$
7 423	6 320	$\frac{7\ 423}{6\ 320} \times 100 = 117.5$	5 432	4 377	$\frac{5\ 432}{4\ 377} \times 100 = 124.1$
8 443	6 388	$\frac{8\ 443}{6\ 388} \times 100 = 132.2$	6 305	4 438	$\frac{6\ 305}{4\ 438} \times 100 = 142.1$
9 837	6 418	$\frac{9\ 837}{6\ 418} \times 100 = 153.3$	7 198	4 460	$\frac{7\ 198}{4\ 460} \times 100 = 161.4$

The index is current-year weighted and is therefore a Paasche index.

6.13 i These statistics are taken from the publication *National Income and Expenditure*. It is the practice to publish National Income Statistics at the prices prevailing during the year concerned and at the prices prevailing at a base date. At current prices the increase in consumer expenditure is made up of two components — an increase in the volume of expenditure and an increase in prices due to inflation. The expenditure revalued at 1970 prices shows the increase in expenditure with the effect of inflation eliminated and hence shows volume changes.

ii

	Current prices	*1970 prices*
	$p_n q_n$	$p_o q_n$
1970	31 387	31 387
1971	34 881	32 241
1972	39 472	34 179
1973	44 855	35 759

Price index: $\frac{\Sigma p_n q_n}{\Sigma p_o q_n} \times 100$ (Paasche) — Quantity index: $\frac{\Sigma p_o q_n}{\Sigma p_o q_o} \times 100$ (Laspeyres)

	Price index	Quantity index
1970	$\frac{31\ 387}{31\ 387} \times 100 = 100$	$\frac{31\ 387}{31\ 387} \times 100 = 100$
1971	$\frac{34\ 881}{32\ 241} \times 100 = 108.2$	$\frac{32\ 241}{31\ 387} \times 100 = 102.7$
1972	$\frac{39\ 472}{34\ 179} \times 100 = 115.5$	$\frac{34\ 179}{31\ 387} \times 100 = 108.9$
1973	$\frac{44\ 855}{35\ 759} \times 100 = 125.4$	$\frac{35\ 759}{31\ 387} \times 100 = 113.9$

6.14

	Average earnings (May 1976 = 100)	*Retail price index* (May 1976 = 100)
1976		
May	$\frac{259.6}{259.6} \times 100 = 100$	$\frac{155.2}{155.2} \times 100 = 100$
June	$\frac{261.2}{259.6} \times 100 = 100.6$	$\frac{156.0}{155.2} \times 100 = 100.5$
July	$\frac{263.1}{259.6} \times 100 = 101.3$	$\frac{156.3}{155.2} \times 100 = 100.7$
August	= 102.9	= 102.1
September	= 102.5	= 103.5
October	= 103.6	= 105.3
November	= 104.9	= 106.8
December	= 106.7	= 108.2
1977		
January	= 107.1	= 111.1
February	= 107.4	= 112.2
March	= 109.3	= 113.3
April	= 109.1	= 116.2
May	= 110.3	= 117.1

6.15

		Earnings (i)	*Retail price index* (ii)	*Earnings deflated* (i) ÷ (ii)
1976	May	100	100	100
	June	100.6	100.5	100.1
	July	101.3	100.7	100.6
	Aug	102.9	102.1	100.8
	Sept	102.5	103.5	99.0
	Oct	103.6	105.3	98.4
	Nov	104.9	106.8	98.2
	Dec	106.7	108.2	98.6

		Earnings (i)	*Retail price index* (ii)	*Earnings deflated* (i) ÷ (ii)
1977	Jan	107.1	111.1	96.4
	Feb	107.4	112.2	95.7
	Mar	109.3	113.3	96.5
	Apr	109.1	116.2	93.9
	May	110.3	117.2	94.1

The earnings-deflated series shows clearly that 'real' earnings fell markedly from September 1976 onwards.

Chapter 7

7.1

Year	*Profits*	*Five year total*	*Five year moving average*
1965	149		
1966	182		
1967	219	955	191.0
1968	200	1 037	207.4
1969	205	1 136	227.2
1970	231	1 229	245.8
1971	281	1330	266.0
1972	312	1 462	292.4
1973	301	1 611	322.2
1974	337	1 669	333.8
1975	380	1 713	342.6
1976	339	1 809	361.8
1977	356		
1978	397		

7.2

Data	*Add in fours*	*Add in pairs*	*Trend* Col (iii) ÷ 8	Col (i) − Col (iv)	*Seasonal variation* (see below)	*Seasonally adjusted data*
Col (i)	Col (ii)	Col (iii)	Col (iv)	Col (v)	Col (vi)	Col (i) − Col (vi)
6						
17	64					
33	64	128	16	+17		
8	66	130	16.25	− 8.25		
6	67	133	16.625	−10.625		
19	69	136	17	+ 2		
34	72	141	17.625	+16.375		
10	74	146	18.25	− 8.25		
9	79	153	19.125	−10.125	−10	19
21	79	158	19.75	+ 1.25	+ 2	19
39					+16	23
10					− 8	18

Year	*Quarter 1*	*Quarter 2*	*Quarter 3*	*Quarter 4*
1973			+17	- 8.25
1974	−10.625	+2	+16.375	- 8.25
1975	−10.125	+1.25		
Total	−20.75	+3.25	+33.375	−16.5
Average	−10.375	+1.625	+16.687 5	- 8.25

As original data correct to nearest integer there is no point in giving seasonal adjustments other than to nearest integer. Seasonal adjustments rounded (16.687 5 rounded down to 16) to ensure that seasonal adjustments total zero — if we had rounded in the usual way the total would have been 1. Seasonal adjustments Quarter 1, −10; Quarter 2, +2; Quarter 3, +16; Quarter 4, −8.

7.3 Multiplicative model: Col (v) = Col (i) ÷ Col (iv) × 100

29				
22	97			
17	101	198	24.75	68.69
29	102	203	25.375	114.29
33	103	205	25.625	128.78
23	107	210	26.25	87.62
18	115	222	27.75	64.86
33	119	234	29.25	112.82
41	123	242	30.25	135.54
27	127	250	31.25	86.40
22	130	257	32.125	68.48
37	131	261	32.625	113.41
44	130	261	32.625	134.87
28				
21				

Year	*Quarter 1*	*Quarter 2*	*Quarter 3*	*Quarter 4*	*Adjustments*
1972			68.69	114.29	133.06
1973	128.78	87.62	64.86	112.82	87.01
1974	135.54	86.40	68.48	113.41	67.34
1975	134.87				113.51
Total	399.19	174.02	202.03	340.52	400.92
Average	133.06	87.01	67.34	113.51	Subtract
Adjusted	132.83	86.78	67.11	113.28	0.23 from each

7.4 A plot of the trend gives a curve, but the last five points are nearly in a straight line. The projection of a straight line from these five points gives trend values of about 21, 22, 23, 23 for the four quarters of 1976. Adding algebraically the seasonal adjustments, the forecasts are: 21 – 10 = 11, 22 + 2 = 24, 23 + 16 = 39, 23 – 8 = 15.

7.5 A plot of the trend yields a curve, thus the projection of the trend is therefore a subjective judgment. A straight line gives a value of about 36 for the fourth quarter of 1975. Multiplying by the seasonal index for the fourth quarter, 113.3, 36 × 113.3 = 41 gives the forecast.

7.6. The components which make up a time series are (a) trend (b) cyclical (c) seasonal (d) residual.

The trend shows the underlying movement of the series. The trend for many economic series is upward, but in some areas such as coal output the trend has been downward. The cyclical movement is usually caused by the effects of booms and slumps and these produce an oscillation about the trend; recently the period of these oscillations has been 3 to 5 years. The seasonal movement is due to the influence of the climate on the time series. Some products, such as ice cream, sell a greater volume in the summer months. The residual components are those factors which affect the time series but are not explained by the 3 components mentioned above. The residual components are usually due to unpredictable factors such as strikes, very unseasonable weather etc.

Additive model

Data Col 1	*Add in fours* Col 2	*Add in pairs* Col 3	*Trend* Col 4	Col 1 – Col 4 = Col 5
40				
37	133			
25	119	252	31.5	–6.5
31	112	231	28.875	+2.125
26	106	218	27.25	–1.25
30	96	202	25.25	+4.75
19	95	191	23.875	–4.875
21	90	185	23.125	–2.125
25	85	175	21.875	+3.125
25	82	167	20.875	+4.125
14				
18				

Year	*Quarter 1*	*Quarter 2*	*Quarter 3*	*Quarter 4*
1973			– 6.5	+2.125
1974	–1.25	+4.75	– 4.875	–2.125
1975	+3.125	+4.125		
Total	+1.875	+8.875	–11.375	0
Average	+0.937 5	+4.437 5	– 5.687 5	0
Adjustment	–0.078 1	–0.078 1	– 0.078 1	–0.078 1
Seasonal adjustments	+1	+5	– 6	0

Sales for 1975 seasonally adjusted are 25 – 1 = 24, 25 – 5 = 20, 14 – (-6) = 20, 18 – 0 = 18.

7.7

Local authorities			*Private*		
128			171		
116			180		
129	652	130.4	178	928	185.6
124	689	137.8	178	974	194.8
155	750	150.0	221	1 003	200.6
165	821	164.2	217	1 029	205.8
177	885	177.0	209	1 077	215.4
200	911	182.2	204	1 042	208.4
188	923	184.6	226	999	199.8
181	901	180.2	186	986	197.2
177	821	164.2	174	983	196.6
155	736	147.2	196	948	189.6
120			201		
103			191		

7.8 i Some data have marked seasonality and thus consecutive values of such data can show wide fluctuations; it is not possible to judge the underlying movement of the series from the actual data. If the data are seasonally adjusted, then the resulting figure(s) represent the trend plus any residual components. An examination of the last few seasonally adjusted values indicates the movement of the underlying trend. For example, unemployment statistics are affected by seasonal factors; the last few months of the unemployment statistics, seasonally adjusted, show the government whether the underlying trend of unemployment is upward, or downward or level. Such information is needed by the government for the management of the economy.

ii **Additive model**

Data Col 1	*Add in fours* Col 2	*Add in pairs* Col 3	*Trend* Col. 4	Col 1 - Col 4 = Col 5
83				
91	400			
95	410	810	101.25	- 6.25
131	423	833	104.125	+26.875
93	440	863	107.875	-14.875
104	463	903	112.875	- 8.875
112	484	947	118.375	- 6.375
154	496	980	122.5	+31.5
114	508	1 004	125.5	-11.5
116	522	1 030	128.75	-12.75
124				
168				

Year	*Quarter 1*	*Quarter 2*	*Quarter 3*	*Quarter 4*
1971			- 6.25	+26.875
1972	-14.875	- 8.875	- 6.375	+31.5
1973	-11.5	-12.75		
Total	-26.375	-21.625	-12.625	+58.375
Average	-13.187 5	-10.812 5	- 6.312 5	+29.187 5
Adjustment	- 0.281 25	- 0.281 25	- 0.281 25	- 0.281 25
Seasonal adjustments	-13	-10	- 6	+29

7.9 **a** The additive model should be used when the amplitude of the seasonal movement is independent of the trend. The multiplicative model should be used when the amplitude of the seasonal movement is proportional to the trend. To find out which model to use in practice a plot of the data and the trend indicates whether the oscillation about the trend is approximately constant or changes with the trend; if the former, use the additive model, if the latter, use the multiplicative model.

b A company needs to have a forecast of future sales in order to ensure that it has an appropriate stock of the commodity to meet future sales. If the firm has too low a level of stock it may not be able to meet demand, and thus lose sales and risk losing the goodwill of its customers. On the other hand, if the firm has too high a level of stock it will have a considerable amount of capital tied up, and because of obsolescence and/or deterioration the stock may rapidly lose value. An accurate forecast of future sales should help the company to avoid such problems.

c **Additive model**

Data Col 1	*Add in fours* Col 2	*Add in pairs* Col 3	*Trend* Col 4	Col 1 - Col 4 = Col 5
455				
393	1 632			
416	1 617	3 249	406.125	+ 9.875
368	1 626	3 243	405.375	-37.375
440	1 625	3 251	406.375	+33.625
402	1 639	3 264	408	- 6
415	1 669	3 308	413.5	+ 1.5
382	1 721	3 390	423.75	-41.75
470	1 741	3 462	432.75	+37.25
454	1 774	3 515	439.375	+14.625
435				
415				

Year	*Quarter 1*	*Quarter 2*	*Quarter 3*	*Quarter 4*
1974		+ 9.875	−37.375	+ 33.625
1975	− 6	+ 1.5	−41.75	+ 37.25
1976	+ 14.625			
Total	+ 8.625	+ 11.375	−79.125	+ 70.875
Average	+ 4.312 5	+ 5.687 5	−39.562 5	+ 35.437 5
Adjustment	+ 1.468 75	+ 1.468 75	+ 1.468 75	+ 1.468 75
Seasonal adjustments	+ 3	+ 4	−41	+ 34

A plot of the trend shows that after the first four values there is a marked upward trend. A straight line can be drawn through these points and the projected trend value corresponding to the fourth quarter is about 460. Adding algebraically the seasonal factor the sales forecast is 460 + 34 = 494.

7.10 Multiplicative model

Data *Col 1*	*Add in fours* *Col 2*	*Add in pairs* *Col 3*	*Trend* *Col 4*	*Col 1/Col 4 ×* *100 = Col 5*
258				
595	2 071			
879	2 143	4 214	526.75	166.872
339	2 248	4 391	548.875	61.762 7
330	2 444	4 692	586.5	56.266
700	2 477	4 921	615.125	113.798
1 075	2 474	4 951	618.875	173.702
372	2 566	5 040	630	59.047 6
327	2 682	5 248	656	49.847 6
792	2 716	5 398	674.75	117.377
1 191	2 743	5 459	682.375	174.537
406	2 841	5 584	698	58.166 2
354	2 974	5 815	726.875	48.701 6
890	3 030	6 004	750.5	118.588
1 324				
462				

Year	*Quarter 1*	*Quarter 2*	*Quarter 3*	*Quarter 4*
1960			166.872	61.762 7
1961	56.266	113.798	173.702	59.047 6
1962	49.847 6	117.377	174.537	58.166 2
1963	48.701 6	118.588		
Total	154.815	349.762	515.112	178.977
Average	51.605 1	116.587	171.704	59.658 8
Adjustment	− 0.1	− 0.1	− 0.1	− 0.1
Seasonal adjustments	51.7	116.7	171.8	59.8

A plot of the trend shows that the trend is upward. If a straight line is drawn and projected forward, the values of the trend for the first two quarters of 1964 are about 800 and 820. Applying the seasonal adjustments we obtain 800 × 51.7 ÷ 100 = 413 for the first quarter and 820 × 116.7 ÷ 100 = 957 for the second quarter.

The actual values were 440 and 946 (thousands). The forecasts were close but not accurate. These particular data have not been adjusted to take into account the fact that Easter is sometimes in the first quarter and sometimes in the second (such adjustments are not considered in this book) — this fact undoubtedly has some effect on the accuracy of the forecasts.

7.11 Additive model

Data *Col 1*	*Add in fours* *Col 2*	*Add in pairs* *Col 3*	*Trend* *Col 4*	*Col 1 – Col 4* *= Col 5*
78				
95	385			
127	389	774	96.75	+30.25
85	382	771	96.375	−11.375
82	382	764	95.5	−13.5
88	380	762	95.25	− 7.25
127	372	752	94	+33
83	369	741	92.625	−9.625
74	359	728	91	−17
85	358	717	89.625	− 4.625
117				
82				

Year	*Quarter 1*	*Quarter 2*	*Quarter 3*	*Quarter 4*
1975			+30.25	−11.375
1976	−13.5	− 7.25	+33	− 9.625
1977	−17	− 4.625		
Total	−30.5	−11.875	+63.25	−21
Average	−15.25	− 5.937 5	+31.625	−10.5
Adjustment	− 0.015 6	− 0.015 6	− 0.015 6	− 0.015 6
Seasonal adjustments	−15	− 6	+31	−10

ii Applying the seasonal factors the seasonally adjusted values for 1977 are 74 − (−15) = 89; 85 − (−6) = 91; 117 − 31 = 86; 82 − (−10) = 92.

iii Forecast methods have been illustrated in other exercises and you should explain in words the process you used. The accuracy of a forecast which depends on the projection of the trend assumes (i) that the past data are of such a nature that it is appropriate to fit a line (or curve) to the data, (ii) that it is reasonable to assume that the trend projected forward will reflect the future trend. From 1957 onwards there was a marked increase in the number of births and this produces some 20 or so years later a marked increase in the number of people of marriageable age; so it is very likely that the downward trend apparent in this data will be replaced at some time after 1977 by an upward trend.

7.12 Additive model

Data *Col 1*	*Add in fours* *Col 2*	*Add in pairs* *Col 3*	*Trend* *Col 4*	*Col 1 - Col 4* *= Col 5*
162				
163	639			
164	632	1 271	158.875	+5.125
150	625	1 257	157.125	-7.125
155	614	1 239	154.875	+0.125
156	604	1 218	152.25	+3.75
153	600	1 204	150.5	+2.5
140	594	1 194	149.25	-9.25
151	588	1 182	147.75	+3.25
150	585	1 173	146.625	+3.375
147				
137				

Year	*Quarter 1*	*Quarter 2*	*Quarter 3*	*Quarter 4*
1974			+5.125	- 7.125
1975	+0.125	+3.75	+2.5	- 9.25
1976	+3.25	+3.375		
Total	+3.375	+7.125	+7.625	-16.375
Average	+1.687 5	+3.562 5	+3.812 5	- 8.187 5
Adjustment	0.218 75	0.218 75	0.218 75	0.218 75
Seasonal adjustments	+1	+3	+4	-8

b A plot of the trend shows that the trend lies on a curve. If the last few values are projected forward trend values of 144 and 143 are found for the first two quarters of 1977, adding the seasonal factors we have forecasts of 145 and 146.

c The forecasting of future births in post-war Britain has proved to be a very difficult problem. In the comment on question 7.11 it was indicated that the number of people of marriageable age would increase after 1977; this in turn is likely to produce an upturn in the number of births. A forecast of the number of future births depends on forecasting the number of future mothers and this can be done with reasonable accuracy for the next 20 years. Forecasting the future average family size of these mothers is a much more difficult exercise.

Chapter 8

8.1 **i** $P(\text{ace}) = \dfrac{\text{Number of aces}}{\text{Number of cards}} = \dfrac{4}{52} = \dfrac{1}{13}$

ii $P(\text{club}) = \dfrac{\text{Number of clubs}}{\text{Number of cards}} = \dfrac{13}{52} = \dfrac{1}{4}$

iii $P(\text{court card}) = \dfrac{\text{Number of court cards}}{\text{Number of cards}} = \dfrac{12}{52} = \dfrac{3}{13}$

8.2 **a** $P(A) = 0.95$, $P(\overline{A}) = 0.05$, $P(B) = 0.96$, $P(\overline{B}) = 0.04$

i $P(AB) = P(A) \times P(B) = 0.95 \times 0.96 = 0.912$

ii 1 - Probability both circuits fail

$= 1 - P(\overline{A}\,\overline{B}) = 1 - P(\overline{A}) + P(\overline{B}) = 1 - 0.05 \times 0.04 = 0.998$

b **i** $\dfrac{16}{20} \times \dfrac{16}{20} \times \dfrac{16}{20} = 0.512$

ii $P(GGD) = \dfrac{16}{20} \times \dfrac{16}{20} \times \dfrac{4}{20} = 0.128.$

Similarly $P(GDG) = 0.128$ and $P(DGG) = 0.128$.

Required probability $= 3 \times 0.128 = 0.384$

8.3 **i** Possible scores (1 1), (1 2), (2 1), (2 2), (3 1), (1 3); no. of scores $= 10 \times 10 = 100$.

$P(\text{score 4 or less}) = \dfrac{\text{Number of possible scores}}{\text{Number of scores}} = \dfrac{6}{100}$

$= 0.06$

ii $P(\text{at least one 2}) = 1 - P(\text{no 2s}) = 1 - 0.9 \times 0.9 = 1 - 0.81 = 0.19$

iii In (i) 3 scores include at least one 2; there is double counting. Required probability $= 0.19 + 0.06 - 0.03 = 0.22$

8.4 **i** $\dfrac{16}{20} \times \dfrac{15}{19} \times \dfrac{14}{18} = \dfrac{28}{57} = 0.491\,2$

ii $3 \times \dfrac{16}{20} \times \dfrac{15}{19} \times \dfrac{4}{18} = \dfrac{8}{19} = 0.421\,1$

8.5 **i** $P(\text{major adjustment/poor part}) = \dfrac{0.07}{0.16} = 0.437\,5$

ii $P(\text{correctly set up/3 good parts}) = \dfrac{0.514\,425}{0.670\,725} = 0.767$

8.5

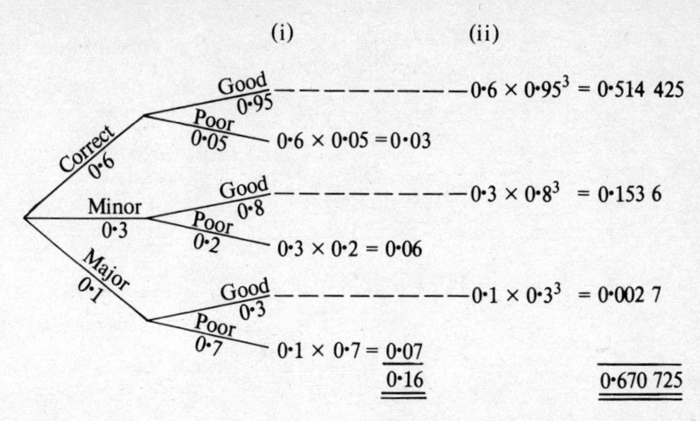

Figure A8.5

8.6 From diagram, number with one defect = 9 + 12 + 13 = 34
Number with at least one defect = 13 + 12 + 9 + 7 + 8 + 4 + 6 = 59
Number with no defects = 100 – 59 = 41

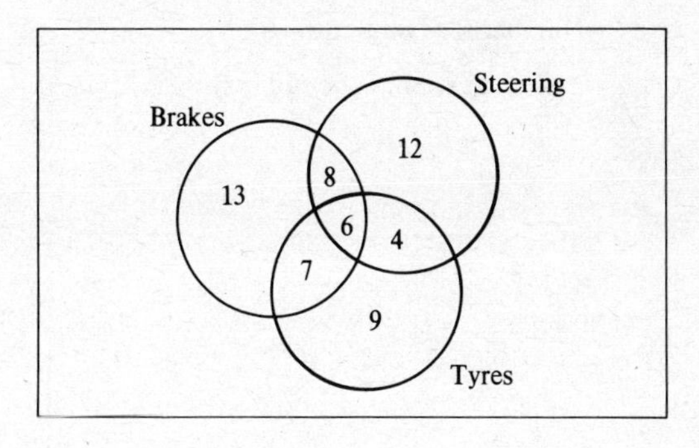

Figure A8.6

8.7 Note that this question does not state whether 'with replacement' or 'without replacement'. The solution assumes 'without replacement'.

a **i** $\frac{10}{16} = 0.625$

ii $\frac{10}{16} \times \frac{9}{15} = 0.375$

iii $\frac{10}{16} \times \frac{9}{15} \times \frac{8}{14} = 0.214\ 3$

b **i** $\frac{10}{16} = 0.625$

ii P(first component defective) × P(second component good)

$= \frac{6}{16} \times \frac{10}{15} = 0.25$

iii $\frac{6}{16} \times \frac{5}{15} \times \frac{4}{14} \times \frac{10}{13} = 0.027\ 5$

8.8 $P(A) = \frac{1}{3}$, $P(\overline{A}) = \frac{2}{3}$, $P(B) = \frac{1}{4}$, $P(\overline{B}) = \frac{3}{4}$

i $P(A) \times P(B) = \frac{1}{3} \times \frac{1}{4} = \frac{1}{12}$.

ii $P(A) + P(B)$ (Mutually exclusive)

$= \frac{1}{3} + \frac{1}{4} = \frac{7}{12}$

iii $(P(\overline{A}))^3 = (\frac{2}{3})^3 = \frac{8}{27}$

iv $1 - P(\text{no sales of B}) = 1 - (P(\overline{B}))^3 = 1 - (\frac{3}{4})^3 = \frac{37}{64}$

8.9 **a** **i** $\frac{12}{30} \times \frac{11}{29} = 0.151\ 7$

ii $2 \times \frac{12}{30} \times \frac{18}{29} = 0.496\ 6$

b Number on course = 9 + 10 + 7 + 4 + 5 + 2 + 3 = 40

Number studying one subject = 9 + 10 + 7 = 26

P(student studies one subject) $= \frac{26}{40} = 0.65$

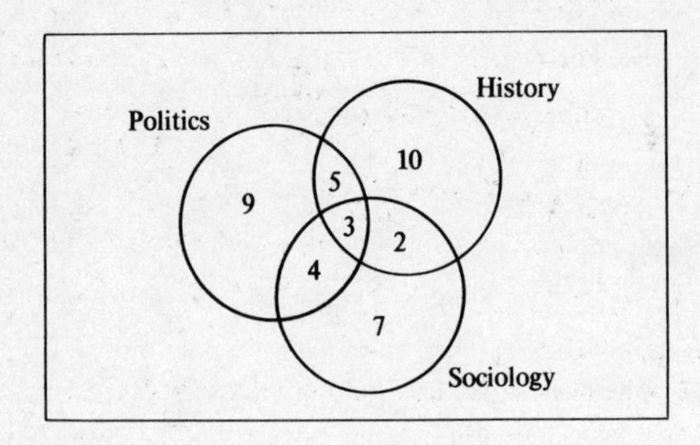

Figure A8.9

8.10 Let components be A, B and C. $P(A) = 0.96$, $P(\overline{A}) = 0.04$, $P(B) = 0.98$, $P(\overline{B}) = 0.02$, $P(C) = 0.99$, $P(\overline{C}) = 0.01$.

a $P(\overline{A}\ \overline{B}\ \overline{C}) = P(\overline{A}) \times P(\overline{B}) \times P(\overline{C}) = 0.04 \times 0.02 \times 0.01 = 0.000\ 008$

b $P(A\overline{B}\,\overline{C}) + P(\overline{A}B\overline{C}) + P(\overline{A}\,\overline{B}C) = (0.96 \times 0.02 \times 0.01) + (0.04 \times 0.98 \times 0.01) + (0.04 \times 0.02 \times 0.99) = 0.001\,376$

c $1 - P(\text{part fails}) = 1 - 0.000\,008 = 0.999\,992$

8.11 Let firms be A, B and C. $P(A) = \frac{1}{2}$, $P(\overline{A}) = \frac{1}{2}$, $P(B) = \frac{1}{3}$, $P(\overline{B}) = \frac{2}{3}$, $P(C) = \frac{1}{4}$, $P(\overline{C}) = \frac{3}{4}$.

a $P(A\overline{B}\,\overline{C}) + P(\overline{A}B\overline{C}) + P(\overline{A}\,\overline{B}C) = (\frac{1}{2} \times \frac{2}{3} \times \frac{3}{4}) + (\frac{1}{2} \times \frac{1}{3} \times \frac{3}{4}) + (\frac{1}{2} \times \frac{2}{3} \times \frac{1}{4}) = \frac{11}{24}$

b $P(\overline{A}\,\overline{B}\,\overline{C}) = \frac{1}{2} \times \frac{2}{3} \times \frac{3}{4} = \frac{1}{4}$

8.12 a At least 4 means 4 or 5 or 6. P(at least 4) $= \frac{1}{6} + \frac{1}{6} + \frac{1}{6} = \frac{1}{2}$

b $1 - P(\text{both dice less than 4}) = 1 - (\frac{1}{2})^2 = \frac{3}{4}$

c $\frac{1}{2} \times \frac{1}{2} = \frac{1}{4}$

d Scores of 3 or less are (1 1), (2 1), (1 2). P(score 3 or less) $= \frac{3}{36} = \frac{1}{12}$. P(sum of scores at least 4) = 1 − P(score 3 or less) $= 1 - \frac{3}{36} = \frac{33}{36} = \frac{11}{12}$

e Differences of 4 are (5 1), (1 5), (2 6), (6 2). P(difference of exactly 4) $= \frac{4}{36} = \frac{1}{9}$. Differences of at least 4, add (6 1), (1 6). P(difference of at least 4) $= \frac{6}{36} = \frac{1}{6}$

8.13 i $\frac{16}{28} \times \frac{15}{27} \times \frac{14}{26} = 0.170\,9$

ii $\frac{17}{28} \times \frac{16}{27} \times \frac{15}{26} = 0.207\,6$

iii $\frac{4}{28} \times \frac{3}{27} \times \frac{2}{26} = 0.001\,221$

iv For a majority of red party members selection must be either 3 red no blue or 2 red 1 blue. P(3 red) $= \frac{11}{28} \times \frac{10}{27} \times \frac{9}{26} = 0.050\,4$. P(2 red 1 blue) $= 3 \times \frac{11}{28} \times \frac{10}{27} \times \frac{17}{26} = 0.2854$. As events are mutually exclusive, required probability $= 0.050\,4 + 0.285\,4 = 0.335\,8$.

8.14 The question infers 'without replacement'.

i $\frac{7}{12} \times \frac{6}{11} = 0.318\,2$

ii $\frac{5}{12} \times \frac{4}{11} = 0.151\,5$

iii $2 \times \frac{7}{12} \times \frac{5}{11} = 0.530\,3$ (alternative method: $1 - 0.318\,2 - 0.151\,5$)

iv As man has twice as much chance of winning if at stage 1 a man and woman are selected, P(man captain) = $\frac{2}{3}$ × 0.530 3 = 0.353 5

v P(woman captain) = $\frac{1}{3}$ × 0.530 3 = 0.176 8

8.15 a

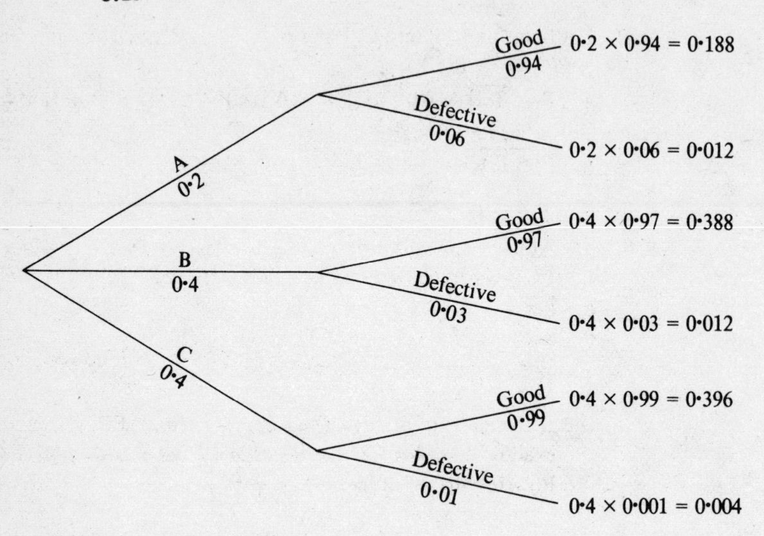

Figure A8.15A

i 0.012 + 0.012 + 0.004 = 0.03 = 3%

ii $\text{P(A/defective part)} = \dfrac{0.012}{0.03} = 0.4$

b

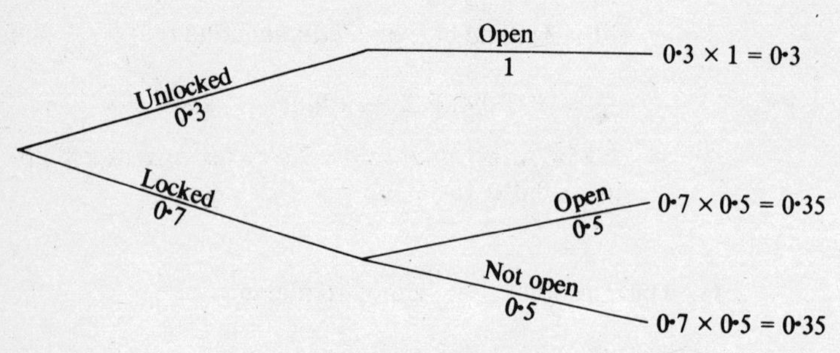

Figure A8.15B

P(open door) = 0.3 + 0.35 = 0.65

8.16 a Percentage who read *Echo* or *News* or both = 20 + 5 + 4 + 3 + 6 + 26 = 64%

b Percentage who read at least one paper = 20 + 5 + 4 + 3 + 6 + 26 + 25 = 89%

c Percentage who read only *Advertiser* = 25%

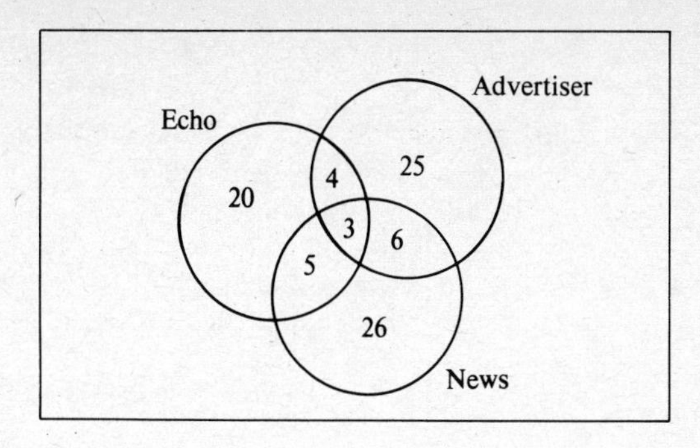

Figure A8.16

8.17

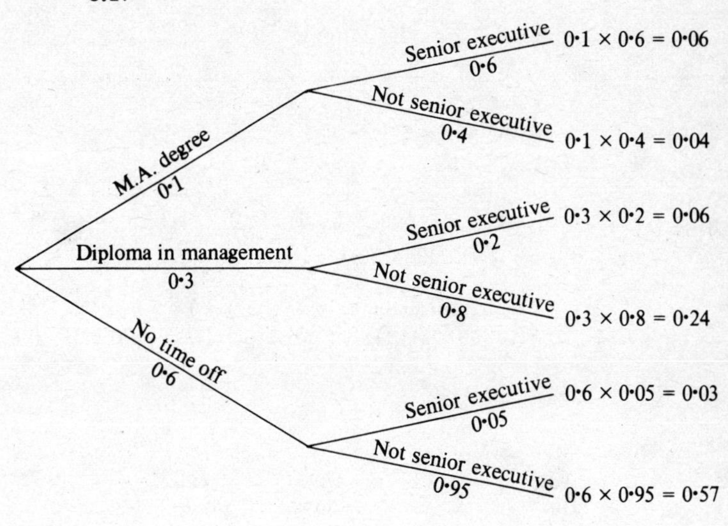

Figure A8.17

a $0.06 + 0.06 + 0.03 = 0.15$

b $$\text{P(Senior Executive/M.A.)} = \frac{0.06}{0.15} = 0.4$$

Chapter 9

9.1 **a** (i) $\Phi(1.3) = 0.903\,2$. (ii) $1 - 0.903\,2 = 0.096\,8$

b (i) $\Phi(0.8) = 0.788\,1$. (ii) $1 - 0.788\,1 = 0.211\,9$

c (i) $\Phi(1.37) = 0.914\,7$. (ii) $1 - 0.914\,7 = 0.085\,3$

9.2 $\Phi(1.88) - \Phi(0.32) = 0.969\ 9 - 0.625\ 5 = 0.344\ 4$

9.3 $\Phi(-0.36) = 1 - \Phi(0.36) = 1 - 0.640\ 6 = 0.359\ 4$;
$\Phi(1.62) - \Phi(-0.36) = 0.947\ 4 - 0.359\ 4 = 0.588\ 0$

9.4 If area between $-x$ and $+x$ is 0.9, each tail is 0.05; we require the value of x such that area to the left of x is $0.9 + 0.05 = 0.95$, $\Phi(x) = 0.95$
$\Phi(1.64) = 0.949\ 5$, $\Phi(1.65) = 0.950\ 5$, thus $x = 1.645$

9.5 $$z = \frac{x - \mu}{\sigma}$$

i $$z = \frac{15 - 10}{4} = 1.25 \quad \Phi(1.25) = 0.894\ 4$$

ii $$z = \frac{12 - 10}{4} = 0.5$$
$1 - \Phi(0.5) = 1 - 0.691\ 5 = 0.308\ 5$

iii $$z = \frac{3 - 10}{4} = -1.75$$
$1 - \Phi(-1.75) = 1 - (1 - \Phi(1.75)) = \Phi(1.75) = 0.959\ 9$

iv $$z_1 = \frac{5 - 10}{4} = -1.25, \qquad z_2 = \frac{9 - 10}{4} = -0.25$$
Required probability $= \Phi(z_2) - \Phi(z_1) = \Phi(-0.25) - \Phi(-1.25)$ $= 1 - \Phi(0.25) - (1 - \Phi(1.25)) = \Phi(1.25) - \Phi(0.25) = 0.894\ 4$ $- 0.598\ 7 = 0.295\ 7$

9.6

x	f	fx	fx^2
90	3	270	24 300
110	13	1 430	157 300
130	20	2 600	338 000
150	35	5 250	787 500
170	16	2 720	462 400
190	10	1 900	361 000
210	3	630	132 300
	100	14 800	2 262 800

$$\bar{x} = \frac{\Sigma fx}{\Sigma f} = \frac{14\ 800}{100} = 148$$

$$\sigma = \sqrt{\frac{\Sigma fx^2}{\Sigma f} - \left(\frac{\Sigma fx}{\Sigma f}\right)^2} = \sqrt{\frac{2\ 262\ 800}{100} - \left(\frac{14\ 800}{100}\right)^2}$$

$= \sqrt{22\ 628 - 21\ 904} = \sqrt{724} = 26.91$

No $= \bar{x} + 1.645 \times 26.91 = 192.3$

Taking the next integer above, 193 need to be held in stock.

9.7 **a** $\Phi(1.2) = 0.884\,9$
b $1 - \Phi(0.7) = 1 - 0.758\,0 = 0.242$
c $\Phi(2.2) - \Phi(1.4) = 0.986\,1 - 0.919\,2 = 0.066\,9$
d $\Phi(-0.6) = 1 - \Phi(0.6) = 1 - 0.725\,7 = 0.274\,3$
e $1 - \Phi(-0.4) = 1 - (1 - \Phi(0.4)) = \Phi(0.4) = 0.655\,4$
f $\Phi(0.2) - \Phi(-0.5) = \Phi(0.2) - (1 - \Phi(0.5)) = 0.579\,3 - (1 - 0.691\,5) = 0.270\,8$

9.8. **a** $z = \dfrac{5-4}{2} = 0.5, \quad \Phi(0.5) = 0.691\,5$
b $z = \dfrac{6-4}{2} = 1, \quad 1 - \Phi(1) = 1 - 0.841\,3 = 0.158\,7$
c $z_1 = \dfrac{7-4}{2} = 1.5, \quad z_2 = \dfrac{5.5-4}{2} = 0.75,$
$\Phi(1.5) - \Phi(0.75) = 0.933\,2 - 0.773\,4 = 0.159\,8$
d $z = \dfrac{2.5-4}{2} = -0.75,$
$\Phi(-0.75) = 1 - \Phi(0.75) = 1 - 0.773\,4 = 0.226\,6$
e $z_1 = \dfrac{4.8-4}{2} = 0.4, \quad z_2 = \dfrac{3.5-4}{2} = -0.25,$
$\Phi(0.4) - \Phi(-0.25) = 0.655\,4 - (1 - 0.598\,7) = 0.254\,1$

9.9 **a** $z = \dfrac{503-500}{10} = 0.3, \quad \Phi(0.3) = 0.617\,9$
b $z = \dfrac{520-500}{10} = 2,$
$1 - \Phi(2) = 1 - 0.977\,25 = 0.022\,75$
c $z_1 = \dfrac{515-500}{10} = 1.5, \quad z_2 = \dfrac{510-500}{10} = 1,$
$\Phi(1.5) - \Phi(1) = 0.933\,2 - 0.841\,3 = 0.091\,9$
d $z = \dfrac{482-500}{10} = -1.8,$
$\Phi(-1.8) = 1 - \Phi(1.8) = 1 - 0.964\,1 = 0.035\,9$

9.10 **a** $z = \dfrac{72-68}{2.5} = 1.6,$
$1 - \Phi(1.6) = 1 - 0.945\,2 = 0.054\,8 = 5.48\%$
b $z = \dfrac{62-68}{2.5} = -2.4,$
$\Phi(-2.4) = 1 - \Phi(2.4) = 1 - 0.991\,8 = 0.008\,2 = 0.82\%$
c $z_1 = \dfrac{70-68}{2.5} = 0.8, \quad z_2 = \dfrac{65-68}{2.5} = -1.2,$
$\Phi(0.8) - \Phi(-1.2) = 0.788\,1 - (1 - 0.884\,9) = 0.673 = 67.3\%$

9.11 We need $\Phi(-x) = 0.01$; thus $1 - \Phi(x) = 0.01$, $\Phi(x) = 0.99$, $x = 2.33$

No. $= 500 + 2.33 \times 5 = 511.65 = 512$ (to nearest gramme)

9.12 a The normal distribution curve is symmetrical and is bell-shaped. The normal curve has been found to fit various types of data — errors in measurement, intelligence quotients of children, heights and weights of animals and people. The normal distribution is very important in statistical theory. It may be shown that in general the distribution of the mean of a random sample (the sampling distribution of the mean) has a normal distribution irrespective of whether the population from which the sample was drawn is normal or not. The proof of this result is based on the 'central limit' theorem, proof of which is to be found in advanced books on statistical theory. In the construction of a confidence interval for the mean (see Chapter 10) we use the fact that the sample mean $\bar{x}$ has a normal distribution with mean μ and standard deviation $\frac{\sigma}{\sqrt{n}}$. Similarly in significance tests based on the mean we use the normal distribution.

In mass production, components are often found to have a length (or width or diameter) which is normally distributed. Using this knowledge, we can estimate the fraction of the components which are expected to lie within certain dimensions. Most quality control procedures are set up using the normal distribution. Weights and measures legislation protects the public against the sale of underweight goods. Manufacturers use the normal distribution to evaluate the average weight of product to pack to minimise the risk of an underweight product.

b $z_1 = \frac{9.96 - 10}{0.025} = -1.6, \quad z_2 = \frac{10.05 - 10}{0.025} = 2$

i correct size $= \Phi(2) - \Phi(-1.6) = 0.977\ 25 - (1 - 0.945\ 2)$
$= 0.922\ 45$

ii oversize $= 1 - \Phi(2) = 1 - 0.977\ 25 = 0.022\ 75$

iii undersize $= \Phi(-1.6) = 1 - \Phi(1.6) = 1 - 0.945\ 2$
$= 0.054\ 8$

Contribution to profit $=$ Revenue $-$ Costs $= 1\ 000 \times 0.922\ 45 \times 18 + 1\ 000 \times 0.022\ 75 \times 18 + 1\ 000 \times 0.054\ 8 \times 1 - (1\ 000 \times 10 + 1\ 000 \times 0.022\ 75 \times 3) = £166.041 + £4.095 + £0.548 - £100 - £0.682\ 5 = £70$

9.13 a Binomial distribution $n = 5$, $p = 0.15$, $q = 0.85$

i $P(0) = q^n = 0.85^5 = 0.443\ 7$

ii $P(1) = npq^{(n-1)} = 5 \times 0.15 \times 0.85^4 = 0.391\ 5$

iii $P(2) = \dfrac{n(n-1)p^2q^{(n-2)}}{2} = \dfrac{5 \times 4 \times 0.15^2 \times 0.85^3}{2}$

$= 0.138\ 2$

$P(r \geqslant 3) = 1 - P(0) - P(1) - P(2) = 1 - 0.443\ 7 - 0.391\ 5 - 0.138\ 2 = 0.026\ 6$

b Poisson distribution $m = 5$

i $P(0) = e^{-m} = e^{-5} = 0.006\ 7$

$P(r \geqslant 2) = 1 - P(0) - P(1)$

$P(1) = me^{-m} = 5 \times e^{-5} = 0.033\ 7.$

$P(r \geqslant 2) = 1 - 0.006\ 7 - 0.033\ 7 = 0.959\ 6$

9.14 Binomial distribution $n = 10$, $p = 0.25$, $q = 0.75$.

i $P(1) = npq^{n-1} = 10 \times 0.25 \times 0.75^9 = 0.187\ 7$

ii P(at least one) $= 1 - P(0)$. $P(0) = q^n = 0.75^{10} = 0.056\ 3$
$1 - 0.056\ 3 = 0.943\ 7$

iii $P(r \geqslant 3) = 1 - P(0) - P(1) - P(2) - P(3)$

$P(2) = \dfrac{n(n-1)p^2q^{n-2}}{2!} = \dfrac{10 \times 9 \times 0.25^2 \times 0.75^8}{2}$

$= 0.281\ 6$

$P(3) = \dfrac{n(n-1)(n-2)p^3q^{n-3}}{3 \times 2}$

$= \dfrac{10 \times 9 \times 8 \times 0.25^3 \times 0.75^7}{3 \times 2}$

$= 0.250\ 3$

$P(r \geqslant 3) = 1 - 0.056\ 3 - 0.187\ 7 - 0.281\ 6 - 0.250\ 3 = 0.224\ 1$

9.15 Poisson distribution $m = 4$. $P(0) = e^{-m} = e^{-4} = 0.018\ 3$
$P(r \geqslant 4) = 1 - P(0) - P(1) - P(2) - P(3) - P(4)$. $P(1) = me^{-m} = 4 \times e^{-4} = 0.073\ 3$

$P(2) = \dfrac{m^2 e^{-m}}{2!} = \dfrac{4^2 \times e^{-4}}{2} = 0.146\ 5$

$P(3) = \dfrac{m^3 e^{-m}}{3!} = \dfrac{4^3 \times e^{-4}}{3 \times 2} = 0.195\ 4$

$P(4) = \dfrac{m^4 e^{-m}}{4!} = \dfrac{4^4 \times e^{-4}}{4 \times 3 \times 2} = 0.195\ 4$

$P(r \geqslant 4) = 1 - 0.018\ 3 - 0.073\ 3 - 0.146\ 5 - 0.195\ 4 - 0.195\ 4 = 0.371$

Expected number of days $= 20 \times 0.371 = 7.4 = 7$ days

9.16 Binomial distribution $n = 100$, $p = 0.8$, $q = 0.2$. With large n it is appropriate to use the normal approximation to the Binomial distribution. $np = 100 \times 0.8 = 80$, $\sqrt{np(1 - p)} = \sqrt{100 \times 0.8 \times 0.2} = 4$. Thus $\mu = 80$, $\sigma = 4$. We need the probability that 87 or more customers appear; we measure from 86.5. $z = \frac{x - \mu}{\sigma} = \frac{86.5 - 80}{4} = 1.625$

Required probability $= 1 - \Phi(1.625) = 1 - 0.948 = 0.052$

Chapter 10

10.1 From 3.2 and 3.5, $\bar{x} = 0.05$, $s = 1.403$, $n = 8\,120$, $\bar{x} + \frac{1.96s}{\sqrt{n}}$ to $\bar{x} - \frac{1.96s}{\sqrt{n}} = 5.05 + \frac{1.96 \times 1.403}{\sqrt{8120}}$ to $5.05 - \frac{1.96 \times 1.403}{\sqrt{8120}}$

$= 5.05 + 0.030\,5$ to $5.05 - 0.030\,5 = 5.08$ to 5.02

10.2 From 3.3 and 3.6, $\bar{x} = 29.4$, $s = 12.52$. Null hypothesis: $\mu = 33\,100$. Alternative hypothesis: $\mu \neq 33\,100$.

$$z = \frac{\bar{x} - \mu}{\frac{s}{\sqrt{n}}} = \frac{29\,400 - 33\,100}{\frac{12\,520}{\sqrt{50}}} = 2.09 \text{ (numerical value)}$$

$z > 1.96$, thus significant at 5%, therefore reject null hypothesis. We conclude that there is some evidence of a difference between the average sales of the representatives of the two companies.

10.3 $p = 0.13$, $n = 1\,000$.

$p + 1.96\sqrt{\frac{p(1-p)}{n}}$ to $p - 1.96\sqrt{\frac{p(1-p)}{n}}$

$= 0.13 + 1.96\sqrt{\frac{0.13\,(1 - 0.13)}{1\,000}}$ to $0.13 - 1.96\sqrt{\frac{0.13\,(1 - 0.13)}{1\,000}}$

$= 0.13 + 0.020\,8$ to $0.13 - 0.020\,8 = 0.15$ to 0.11.

To obtain percentages multiply by 100: 15% to 11%.

10.4 $p = \frac{307}{580} = 0.529$. Null hypothesis: $\pi = 0.48$. Alternative hypothesis: $\pi \neq 0.48$. Thus a one-tailed test: $z = \frac{p - \pi}{\sqrt{\frac{\pi(1-\pi)}{n}}}$

$$= \frac{0.529 - 0.48}{\sqrt{\frac{0.48(1 - 0.48)}{580}}} = 2.38$$

$z > 2.33$, the 1% point one-tailed. We therefore reject the null hypothesis and conclude that there is good evidence that the percentage of households with a telephone has increased.

10.5 Old system $n_o = 40, \bar{x}_o = 18.2, s_0 = 0.6$. New system $n_n = 40, \bar{x}_n = 17.8, s_n = 0.5$. Null hypothesis: $\mu_o = \mu_n$. Alternative hypothesis: $\mu_o > \mu_n$. One-tailed test.

$$z = \frac{\bar{x}_o - \bar{x}_n}{\sqrt{\frac{s_o^2}{n_o} + \frac{s_n^2}{n_n}}} = \frac{18.2 - 17.8}{\sqrt{\frac{0.6^2}{40} + \frac{0.5^2}{40}}} = 3.24.$$

$z > 2.88$, the 0.2% point on a one-tailed test.
We therefore reject the null hypothesis and conclude that there is strong evidence that the new system has reduced invoice errors.

10.6 $p_1 = 0.4, p_2 = 0.34$ (this assumes that those not opposed are in favour of the proposal; 0.34 may include a proportion of 'don't knows').
Null hypothesis: $\pi_1 = \pi_2$. Alternative hypothesis: $\pi_1 \neq \pi_2$. Two-tailed test.

$$z = \frac{p_1 - p_2}{\sqrt{p(1-p)\left(\frac{1}{n_1} + \frac{1}{n_2}\right)}} \text{ where } p = \frac{n_1p_1 + n_2p_2}{n_1 + n_2}$$

$$= \frac{600 \times 0.4 + 1\,200 \times 0.34}{600 + 1\,200} = 0.36$$

$$z = \frac{0.4 - 0.34}{\sqrt{0.36 \times 0.64\left(\frac{1}{600} + \frac{1}{1\,200}\right)}} = 2.5$$

$z > 1.96$, thus significant at 5%. We therefore reject the null hypothesis and conclude that there is some evidence that the findings are not consistent with each other.

10.7 $\bar{x} = 48.50, s = 6.50, n = 200$

$$95\%: 48.50 + \frac{1.96 \times 6.5}{\sqrt{200}} \text{ to } 48.50 - \frac{1.96 \times 6.5}{\sqrt{200}}$$

$= 48.50 + 0.901$ to $48.50 - 0.901$
$= 49.40$ to 47.60

$$99\%: 48.50 + \frac{2.58 \times 6.5}{\sqrt{200}} \text{ to } 48.50 - \frac{2.58 \times 6.5}{\sqrt{200}}$$

$= 48.50 + 1.186$ to $48.50 - 1.186$
$= 49.68$ to 47.31

10.8 b

x	f	fx	fx^2
2.5	45	112.5	281.25
7.5	30	225.0	1 687.5
12.5	11	137.5	1 718.75
17.5	9	157.5	2 756.25
22.5	2	45.0	1 012.5
27.5	1	27.5	756.25
32.5	1	32.5	1 056.25
37.5	1	37.5	1 406.25
	100	775.0	10 675.00

$$\bar{x} = \frac{\Sigma fx}{\Sigma f} = \frac{775}{100} = 7.75$$

$$s = \sqrt{\frac{\Sigma fx^2}{\Sigma f} - \left(\frac{\Sigma fx}{\Sigma f}\right)^2} = \sqrt{\frac{10\,675}{100} - \left(\frac{775}{100}\right)^2}$$

$$= \sqrt{46.687\,5} = 6.83$$

Null hypothesis: $\mu = 7.4$

Alternative hypothesis: $\mu \neq 7.4$; two-tailed test

$$z = \frac{7.75 - 7.4}{\frac{6.83}{\sqrt{2\,844}}} = 2.73$$

$z > 2.58$, thus significant at 1%, therefore reject null hypothesis. We conclude that there is good evidence that the average length of residence of mortgagors has changed.

c The test we have employed is based on the fact that when we take a random sample and calculate the sample mean we know that the sampling distribution of the sample mean is a normal distribution with mean μ and standard deviation $\frac{\sigma}{\sqrt{n}}$.

We also take into account the fact that the sample, though drawn at random, may or may not be representative of the population from which the sample was drawn.

In part (b) of this question we are testing whether the average length of residence of mortgagors in 1974 ($= \mu$) has changed from the 1971 figure of 7.4. We set up the null hypothesis that $\mu = 7.4$, i.e. that the length of residence has not changed; and the alternative hypothesis that $\mu \neq 7.4$, i.e. that the length of residence has changed. Now $\bar{x}$, the sample mean, is our estimate of μ. If $\bar{x}$ differs from 7.4, this can be either because

i the average length of residence has changed from 7.4; or

ii the average length of residence has not changed but the sample was not representative of the population.

The fact that $\bar{x}$ has a normal distribution enables us to work out the probability that the difference between $\bar{x}$ and 7.4 is due to (ii). If this probability is above a certain level —

above 0.05 by convention — we agree that (ii) is true and we accept the null hypothesis. If the probability is less than 0.05 we accept that (ii) is not true and conclude that (i) is true.

To find the probability we use the normal distribution tables. We divide $(\bar{x} - 7.4)$ by the standard deviation of $\bar{x}$, $\frac{s}{\sqrt{n}}$, i.e. $z = \frac{\bar{x} - 7.4}{\frac{s}{\sqrt{n}}}$ Now z has the the normal distribution with mean 0 and standard deviation 1. From the normal tables we know that if $z = 1.96$ we are at the 5% point; thus if z is less than 1.96 then (ii) is likely to be true; if z is greater than 1.96 then (i) is likely to be true.

10.9 Null hypothesis: $\mu = 100$. Alternative hypothesis: $\mu \neq 100$. Two-tailed test. $\bar{x} = 103.2$, $s = 15$, $n = 50$.

$$z = \frac{103.2 - 100}{\frac{15}{\sqrt{50}}} = 1.51.$$

$z < 1.96$, therefore accept the null hypothesis. There is no evidence that Polytechnic graduates differ from the standard score for graduates.

10.10 Null hypothesis: $\pi = 0.15$. Alternative hypothesis: $\pi \neq 0.15$. Two-tailed test. $p = 0.2$, $n = 100$.

$$z = \frac{0.2 - 0.15}{\sqrt{\frac{0.15(1 - 0.15)}{100}}} = 1.40.$$

$z < 1.96$, which is not significant, therefore accept the null hypothesis. Confidence interval:

$$0.2 + 1.96\sqrt{\frac{0.2(1 - 0.2)}{100}} \text{ to } 0.2 - 1.96\sqrt{\frac{0.2(1 - 0.2)}{100}}$$

$= 0.2 + 0.078$ to $0.2 - 0.078 = 0.278$ to 0.122

10.11 a

x	f	fx	fx^2
5	59	295	1 475
7	30	210	1 470
9	33	297	2 673
11	22	242	2 662
13	10	130	1 690
15	2	30	450
17	2	34	578
19	3	57	1 083
25	7	175	4 375
	168	1 470	16 456

$$\bar{x} = \frac{\Sigma fx}{\Sigma f} = \frac{1\,470}{168} = 8.75$$

$$s = \sqrt{\frac{\Sigma fx^2}{\Sigma f} - \left(\frac{\Sigma fx}{\Sigma f}\right)^2} = \sqrt{\frac{16\,456}{168} - \left(\frac{1\,470}{168}\right)^2}$$

$$= \sqrt{21.389\,9} = 4.625$$

b Null hypothesis: $\mu_w = \mu_s$
Alternative hypothesis $\mu_w \neq \mu_s$, two-tailed test.

$$z = \frac{\bar{x}_w - \bar{x}_s}{\sqrt{\frac{s_w^2}{n_w} + \frac{s_s^2}{n_s}}} = \frac{8.75 - 7.9}{\sqrt{\frac{4.625^2}{168} + \frac{4.6^2}{422}}} = 2.02$$

$z > 1.96$, thus significant at 5%. We therefore reject the null hypothesis and conclude that there is evidence of a difference between the earnings of women in the West Midlands Region and the South East Region.

c In 10.8(c) we discussed the reasoning of the test employed. Similarly in this case we can argue that the difference between the sample means $\bar{x}_w$ and $\bar{x}_s$ is because either

i there is a difference between the mean earnings of the women in the two regions; or

ii there is no difference between the mean earnings of the women in the two regions, but the difference in the sample means was due to the fact that the sample(s) were not representative of the population.

If we find that z is less than 1.96, we say that the difference is 'not significant', i.e. the difference between sample means could be due to chance effects of sampling. If z is greater than 1.96, we say that the difference is 'significant', i.e. that it is most likely that the difference in sample means is due to the fact that in the two regions earnings do in fact differ.

10.12 a $p = \frac{20}{250} = 0.08.$

95% confidence interval

$$= 0.08 + 1.96\sqrt{\frac{0.08 \times 0.92}{250}} \text{ to } 0.08 - 1.96\sqrt{\frac{0.08 \times 0.92}{250}}$$

$$= 0.08 + 0.034 \text{ to } 0.08 - 0.034 = 0.114 \text{ to } 0.046$$

b Width of confidence interval $= 2 \times 1.96\sqrt{\frac{p(1-p)}{n}}$

$$= 3.92\sqrt{\frac{0.08 \times 0.92}{n}}$$

Thus $0.05 > 3.92\sqrt{\frac{0.08 \times 0.92}{n}}$. Square both sides:

$$0.0025 > 15.366\,4 \times \frac{0.08 \times 0.92}{n}$$

Hence $n > \frac{15.366\,4 \times 0.08 \times 0.92}{0.002\,5}$, $n > 452.4$. Since n must be an integer, $n = 453$

c Null hypothesis: $\pi = 0.04$. Alternative hypothesis: $\pi \neq 0.04$. Two-tailed test.

$$z = \frac{0.08 - 0.04}{\sqrt{\frac{0.04 \times 0.96}{250}}} = 3.23$$

$z > 3.09$, thus significant at 0.2%. Very strong evidence to reject the null hypothesis; we conclude therefore that the error rate has changed.

10.13 a When we select a random sample we have to recognise that the sample, though drawn at random, may or may not be representative of the population. In the example of this question where the 710 households were selected at random, it could be that by chance we chose a higher (or lower) proportion of households in the sample with a telephone than the actual proportion if we considered all households in Wales. Thus the sample proportion could be an over-estimate or an under-estimate of the actual proportion. We say that the sample proportion has a sampling distribution. We may show that in general this sampling distribution is a normal distribution with mean π and standard deviation $\sqrt{\frac{\pi(1-\pi)}{n}}$. We can therefore use the properties of the normal distribution. If we are finding a 95% confidence interval, we can use the fact that 95% of observations are within 1.96 standard deviations of the mean.

From sample data we estimate π by p, the sample proportion. An estimate of the standard deviation is $\sqrt{\frac{p(1-p)}{n}}$.

To incorporate the possibility that a sample proportion could be an under-estimate or an over-estimate of the population proportion we say that we are 95% confident that π lies between $p \pm 1.96\sqrt{\frac{p(1-p)}{n}}$.

b $$\text{p} = \frac{224}{710} = 0.3155$$

$$0.3155 + 1.96\sqrt{\frac{0.3155 \times 0.6845}{710}}$$

$$\text{to } 0.3155 - 1.96\sqrt{\frac{0.3155 \times 0.6845}{710}}$$

$= 0.3155 + 0.0342$ to $0.3155 - 0.0342 = 0.35$ to 0.281; as a percentage 35% to 28.1%.

c Null hypothesis: $\pi_w = \pi_s$. Alternative hypothesis: $\pi_w \neq \pi_s$. Two-tailed test.

$$p_s = \frac{609}{1310} = 0.464\,9.$$

$$p = \frac{224 + 609}{710 + 1\,310} = 0.412\,4$$

$$z = \frac{0.464\,9 - 0.315\,5}{\sqrt{0.412\,4 \times 0.587\,6\left(\frac{1}{1310} + \frac{1}{710}\right)}}$$

$z = 6.51$. $z > 3.09$, thus significant at 0.2%. Very strong evidence to reject null hypothesis. We conclude that there is a difference in the proportion of households in Scotland and Wales possessing a telephone.

10.14

x	f	fx	fx^2
1	18	18	18
2	32	64	128
3	20	60	180
4	19	76	304
5	7	35	175
7	4	28	196
	100	281	1 001

$$\bar{x} = \frac{\Sigma fx}{\Sigma f} = \frac{281}{100} = 2.81$$

$$s = \sqrt{\frac{\Sigma fx^2}{\Sigma f} - \left(\frac{\Sigma fx}{\Sigma f}\right)^2} = \sqrt{\frac{1\,001}{100} - \left(\frac{281}{100}\right)^2}$$

$$= \sqrt{2.113\,9} = 1.45$$

95% confidence interval $= 2.81 + 1.96 \times \frac{1.45}{\sqrt{445}}$

to $2.81 - 1.96 \times \frac{1.45}{\sqrt{445}} = 2.81 + 0.13$ to $2.81 - 0.13$

$= 2.94$ to 2.68

10.15 i Null hypothesis: $\pi = 0.08$. Alternative hypothesis: $\pi \neq 0.08$. Two-tailed test.

$$p = \frac{19}{200} = 0.095,$$

$$z = \frac{0.095 - 0.08}{\sqrt{\frac{0.08 \times 0.92}{200}}} = 0.782.$$

$z < 1.96$. Accept null hypothesis. We conclude that left-handedness does not affect typing ability.

ii $0.8 + 1.96\sqrt{\frac{0.2 \times 0.8}{64}}$ to $0.8 - 1.96\sqrt{\frac{0.2 \times 0.8}{64}}$

$= 0.8 + 0.098$ to $0.8 - 0.098 = 0.898$ to 0.702

This is a 95% confidence interval for the proportion expected to pass. The number expected to pass is found by multiplying these proportions by 64, i.e. 57 to 45.

Chapter 11

11.1 a Null hypothesis: no relationship between the incidence of thumb-sucking and social class.

Observed values				*Expected values*				*Observed – Expected*			
33	49	38	120	22	55	43	120	+11	−6	−5	0
99	281	220	600	110	275	215	600	−11	+6	+5	0
132	330	258	720	132	330	258	720	0	0	0	0

$$\sum\frac{(O-E)^2}{E} = \frac{(+11)^2}{22} + \frac{(-6)^2}{55} + \frac{(-5)^2}{43} + \frac{(-11)^2}{110} + \frac{(+6)^2}{275} + \frac{(+5)^2}{215}$$

$$= 5.5 + 0.6545 + 0.5814 + 1.1 + 0.130\,9 + 0.116\,3 = 8.083$$

Degrees of freedom $(r-1)(c-1) = (2-1)(3-1) = 1 \times 2 = 2$
$\chi^2 = 7.82$ at 2% point. The value of χ^2 at 8.083 is greater than 2% point; we therefore reject the null hypothesis. We conclude therefore that there is some relationship between thumb-sucking and social class.

b See 11.6.

11.2 Null hypothesis: no relationship between residence of parents and towns.

Observed values			*Expected values*			*Observed – Expected*		
64	199	263	76.73	186.27	263	−12.73	+12.73	0
88	170	258	75.27	182.73	258	+12.73	−12.73	0
152	369	521	152	369	521	0	0	0

$$\sum\frac{(O-E)^2}{E} = \frac{(-12.73)^2}{76.73} + \frac{(+12.73)^2}{186.27} + \frac{(+12.73)^2}{75.27} + \frac{(-12.73)^2}{182.73}$$

$$= 2.112 + 0.870 + 2.153 + 0.887 = 6.022$$

Degrees of freedom $(r-1)(c-1) = (2-1)(2-1) = 1 \times 1 = 1$
$\chi^2 = 5.41$ at 2%. The value of $\chi^2 = 6.022$ is greater than 2% point. We therefore reject the null hypothesis. We conclude that there is some evidence that proportionately more couples live near their parents in Bethnal Green than in Woodford.

This particular question could have been tackled by the method of Worked Example No. 10.6 (and conversely Worked Example No. 10.6 could have been solved using χ^2 and in fact many students prefer to do this). If this example had been tackled by the method of Worked Example No. 10.6 we would have employed a one-tailed test; we would have found that z = 2.45 and that the result was significant at 1%. We found using a χ^2 test that the result was significant at 2%. The χ^2 tables are in effect two-tailed. Taking into account the two-tailed nature of χ^2 tables, the result was, therefore, significant at 1%.

11.3 Null hypothesis: no relationship between occupational category and promotional aspirations.

Observed values			*Expected values*			*Observed – Expected*		
47	7	54	30.72	23.28	54	+16.28	−16.28	0
49	30	79	44.94	34.06	79	+ 4.06	− 4.06	0
65	85	150	85.34	64.66	150	−20.34	+20.34	0
161	122	283	161	122	283	0	0	0

$$\sum\frac{(O-E)^2}{E} = \frac{(+16.28)^2}{30.72} + \frac{(-16.28)^2}{23.28} + \frac{(+4.06)^2}{44.94} + \frac{(-4.06)^2}{34.06} + \frac{(-20.34)^2}{85.34} + \frac{(+20.34)^2}{64.66}$$
$$= 8.63 + 11.38 + 0.37 + 0.48 + 4.85 + 6.40$$
$$= 32.11$$

Degrees of freedom $(r-1)(c-1) = (3-1)(2-1) = 2 \times 1 = 2$. $\chi^2 = 9.21$ at 1% point. The value of $\chi^2 = 32.11$ is very much greater than the 1% point. There is therefore very strong evidence to reject the null hypothesis. We conclude therefore that occupational category is related to promotion aspirations.

11.4 Null hypothesis: no relationship between job classification and preference for pension plans.

Observed values			*Expected values*			*Observed – Expected*		
19	23	42	16.8	25.2	42	+2.2	−2.2	0
101	157	258	103.2	154.8	258	−2.2	+2.2	0
120	180	300	120	180	300	0	0	0

$$\sum\frac{(O-E)^2}{E} = \frac{(+2.2)^2}{16.8} + \frac{(-2.2)^2}{25.2} + \frac{(-2.2)^2}{103.2} + \frac{(+2.2)^2}{154.8}$$
$$= 0.288 + 0.192 + 0.047 + 0.031 = 0.558$$

Degrees of freedom $(r-1)(c-1) = (2-1)(2-1) = 1 \times 1 = 1$
$\chi^2 = 3.84$ at 5% point. The value of $\chi^2 = 0.558$ is very much less than the 5% point. We therefore accept the null hypothesis. We conclude that there is no evidence that preference for the two pension plans is related to job classification.

11.5 Null hypothesis: no relationship between sex and job satisfaction.

Observed values

4 204	3 545	1 157	8 906
3 430	1 757	463	5 650
7 634	5 302	1 620	14 556

Expected values

4 670.8	3 244	991.2	8 906
2 963.2	2 058	628.8	5 650
7 634.0	5 302	1 620.0	14 556

Observed – Expected

– 466.8	+301	+165.8	0
+466.8	-301	-165.8	0
0	0	0	0

$$\sum \frac{(O - E)^2}{E} = \frac{(-466.8)^2}{4670.8} + \frac{(301)^2}{3\,244} + \frac{(165.8)^2}{991.2} + \frac{(+466.8)^2}{2963.2} + \frac{(-301)^2}{2\,058} + \frac{(-165.8)^2}{628.8}$$
$$= 46.66 + 27.93 + 27.73 + 73.54 + 44.02 + 43.72 = 263.6$$

Degrees of freedom $(r - 1)(c - 1) = (2 - 1)(3 - 1) = 1 \times 2 = 2$
$\chi^2 = 9.21$ at 1% point. The value of $\chi^2 = 263.6$ is very much greater than the 1% point. We therefore reject the null hypothesis. We conclude that there is very strong evidence that there is a relationship between sex and job satisfaction.

11.6 a The χ^2 test examines sample data to see whether it is possible to draw inferences about the population on the basis of the sample data. When we have data from a sample it must be recognised that the sample, though random, may not represent the population. In this exercise the sample of manual workers may not be representative of all manual workers. In random sampling there is always the risk of obtaining unrepresentative samples; we use the theory of probability to evaluate these risks.

In a χ^2 test we set up the null hypothesis that there is no relationship between the variables concerned — in this exercise the variables are occupation and overcrowding. We work out the expected values under the null hypothesis; the method of doing this is illustrated numerically in part (b). The expected values are compared with the observed values. If the null hypothesis is true, then it is likely that these two values are close together (it is unlikely that the two values are exactly equal, because there is always some variation caused by sampling). If the null hypothesis is not true, then we would expect the observed and expected values to differ. Occasionally observed and expected values differ because the sample is by chance unrepresentative of the population rather than because the null hypothesis is not true. It is therefore necessary to employ a statistical test to evaluate the chance that the sample was unrepresentative of the population. If O = observed value and E = expected, we calculate the value of $\sum \frac{(O - E)^2}{E} = C$ (say). If O and E are close together, then C would be small; if O and E are not close together then C would be large. It is possible to show that under the null hypothesis $C = \sum \frac{(O - E)^2}{E}$ has approximately the χ^2 distribution with $(r - 1)(c - 1)$ degrees of freedom. It is possible to show that if the null hypothesis is true we would expect C to equal $(r - 1)(c - 1)$. If the value of C is much larger than $(r - 1)(c - 1)$, we would doubt whether the null hypothesis is true. The tables of the χ^2 distribution enable us to express this doubt in probability terms; for example, with 2 degrees of freedom $\sum \frac{(O - E)^2}{E} = 5.99$ would indicate that there are 5 chances in 100 (since 5.99 is at the 5% point for two degrees of freedom) that the null hypothesis is true, but the relatively large value of C was obtained because the sample was by chance unrepresentative of the population, and 95 chances in 100 that the null hypothesis is not true. Thus the value of $C = \sum \frac{(O - E)^2}{E}$ when compared with the χ^2 table enables us to determine with a given level of probability whether the null hypothesis is acceptable. The usual convention is to accept the null hypothesis if the value of $C = \sum \frac{(O - E)^2}{E}$ is less than the corresponding 5% point and reject the null hypothesis if the value exceeds the 5% point.

b Null hypothesis: no relationship between occupation and overcrowding as measured by the bedroom standard.

Observed values

150	1 220	3 030	4 400
510	2 410	3 680	6 600
660	3 630	6 710	11 000

Expected values

264	1 452	2 684	4 400
396	2 178	4 026	6 600
660	3 630	6 710	11 000

Observed - Expected

-114	-232	+346	0
+114	+232	-346	0
0	0	0	0

$$\sum \frac{(O-E)^2}{E} = \frac{(-114)^2}{264} + \frac{(-232)^2}{1452} + \frac{(+346)^2}{2684} + \frac{(+114)^2}{396} + \frac{(+232)^2}{2178} + \frac{(-346)^2}{4026}$$

$$= 49.23 + 37.07 + 44.60 + 32.82 + 24.71 + 29.74 = 218.2$$

Degrees of freedom $(r-1)(c-1) = (2-1)(3-1) = 1 \times 2 = 2$
$\chi^2 = 9.21$ at 1% point. The value of $\chi^2 = 218.2$ is very much greater than the 1% point. We therefore reject the null hypothesis and conclude that there is very strong evidence of a relationship between occupation and overcrowding.

11.7 Null hypothesis: no difference in pass rates between the three subjects.

Observed values			*Expected values*			*Observed - Expected*		
324	447	771	303.49	467.51	771	+20.51	-20.51	0
81	211	292	114.94	177.06	292	-33.94	+33.94	0
65	66	131	51.57	79.43	131	+13.43	-13.43	0
470	724	1 194	470.00	724.00	1 194	0	0	0

$$\sum \frac{(O-E)^2}{E} = \frac{(+20.51)^2}{303.49} + \frac{(-20.51)^2}{467.51} + \frac{(-33.94)^2}{114.94} + \frac{(+33.94)^2}{177.06} + \frac{(+13.43)^2}{51.57} + \frac{(-13.43)^2}{79.43}$$

$$= 1.386 + 0.900 + 10.022 + 6.506 + 3.497 + 2.271 = 24.58$$

Degrees of freedom $= (r-1)(c-1) = (3-1)(2-1) = 2 \times 1 = 2$
$\chi^2 = 9.21$ at 1% point. The value of $\chi^2 = 24.58$ is greater than the 1% point. We therefore reject the null hypothesis and conclude that there is very strong evidence of a difference in pass rates between subjects.

An examination of the components which make up the value of $\chi^2 = 24.59$ shows that most of the contribution is from physics. Thus the main factor in making the decision is the markedly different pass rate for physics.

11.8 When data are presented in the form of this exercise, the first step is to obtain the number who were not absent for each region.

Null hypothesis: no relationship between absence from work owing to illness or accident and geographical region.

Observed values			*Expected values*			*Observed – Expected*		
39	948	987	46.70	940.30	987	– 7.70	+ 7.70	0
22	623	645	30.52	614.48	645	– 8.52	+ 8.52	0
49	644	693	32.79	660.21	693	+ 16.21	–16.21	0
110	2 215	2 325	110.00	2 215.00	2 325	– 0.01	+ 0.01	0

$$\sum \frac{(O-E)^2}{E} = \frac{(-7.70)^2}{46.70} + \frac{(+7.70)^2}{940.30} + \frac{(-8.52)^2}{30.52} + \frac{(+8.52)^2}{614.48} + \frac{(+16.21)^2}{32.79} + \frac{(-16.21)^2}{660.21}$$

$$= 1.269 + 0.006 + 2.377 + 0.118 + 8.017 + 0.398 = 12.24$$

Degrees of freedom $(r-1)(c-1) = (3-1)(2-1) = 2 \times 1 = 2$. $\chi^2 = 9.21$ at 1% point. The value of $\chi^2 = 12.24$ is greater than the 1% point. We therefore reject the null hypothesis, and conclude that there is good evidence that there is a relationship between absence from work owing to illness or accident and geographical region.

11.9 Null hypothesis: no relationship between ownership of a television set and social class.

Observed values			*Expected values*			*Observed – Expected*		
15	5	20	15.00	5.00	20	0	0	0
38	7	45	33.75	11.25	45	+4.25	–4.25	0
37	18	55	41.25	13.75	55	–4.25	+4.25	0
90	30	120	90.00	30.00	120	0	0	0

$$\sum \frac{(O-E)^2}{E} = \frac{0}{15} + \frac{0}{5} + \frac{(+4.25)^2}{33.75} + \frac{(-4.25)^2}{11.25} + \frac{(-4.25)^2}{41.25} + \frac{(+4.25)^2}{13.75}$$

$$= 0 + 0 + 0.535 + 1.606 + 0.438 + 1.314$$

$$= 3.893$$

Degrees of freedom $(r-1)(c-1) = (3-1)(2-1) = 2 \times 1 = 2$
$\chi^2 = 5.99$ at 5% point. The value of χ^2 is 3.893. This is less than the 5% point. We therefore accept the null hypothesis, and conclude that there is no evidence of an association between ownership of a television set and social class.

11.10

x	f	fx
0	35	0
1	40	40
2	18	36
3	4	12
4	3	12
	100	90

$$\bar{x} = \frac{\Sigma fx}{\Sigma f} = \frac{90}{100} = 0.9$$

Null hypothesis: Poisson distribution a good fit to the data.
Expected frequencies under null hypothesis = 100 × P(r)

$$100 \times P(0) = 100 \times e^{-m} = 100 \times e^{-0.9} = 40.7$$
$$100 \times P(1) = 100 \times me^{-m} = 100 \times 0.9 \times e^{-0.9} = 36.6$$
$$100 \times P(2) = 100 \times \frac{m^2e^{-m}}{2} = 100 \times \frac{0.9^2 \times e^{-0.9}}{2} = 16.5$$

O	E	O - E
35	40.7	-5.7
40	36.6	+3.4
18	16.5	+1.5
7	6.2	+0.8
100	100	0

$$\chi^2 = \sum \frac{(O - E)^2}{E} = \frac{(-5.7)^2}{40.7} + \frac{(3.4)^2}{36.6} + \frac{(1.5)^2}{16.5} + \frac{(0.8)^2}{6.2}$$
$$= 0.798 + 0.316 + 0.136 + 0.103$$
$$= 1.353$$

Degrees of freedom = 4 – 1 – 1 = 2. x^2 (5%) = 5.99. The value of χ^2 is less than 5.99; therefore result not significant; we accept null hypothesis and conclude that the Poisson distribution is a good fit to the data.

11.11 a i *Tests of association using contingency tables.* These involve investigating the relationship between two variables, at least one of which cannot be quantified. Often used by sociologists when investigating social class differences etc. In calculating the degrees of freedom for $r \times c$ (r rows, c columns) we note that in calculating the expected values, the last row and column can be found by subtracting from the row and column totals. We thus have $(r - 1)(c - 1)$ degrees of freedom.

ii *Goodness of fit tests.* If data occur at random, it is possible to show that a Poisson distribution should fit the data. Thus if it is thought that (say) accidents occur at random, then the Poisson distribution should fit accident data. The goodness of fit test examines whether a Poisson distribution is a good fit to the data. If the

Poisson distribution is a good fit, then we could reasonably conclude that the data are random. If the Poisson distribution is not a good fit, then this would suggest that the data are non-random. In calculating degrees of freedom we take the number of groups used in calculating χ^2 and deduct a degree of freedom for the constraint imposed by the fact that the observed and expected totals are the same; and a further degree of freedom if we calculate a mean or a proportion in order to fit the distribution. In the case of fitting a normal distribution we deduct two degrees of freedom, since we have to find a mean and standard deviation to fit a normal distribution.

b Null hypothesis: shape of bottle has no effect on sales.

The total sales are 611. If the null hypothesis is true, we would expect equal numbers of each shape to be sold. 611 ÷ 3 = 203.67.

Observed	261	149	201
Expected	203.67	203.67	203.67
O – E	+57.33	–54.67	–2.67

$$\chi^2 = \sum \frac{(O-E)^2}{E} = \frac{(57.33)^2}{203.67} + \frac{(-54.67)^2}{203.67} + \frac{(-2.67)^2}{203.67}$$

$$= 16.14 + 14.67 + 0.04 = 30.85$$

Degrees of freedom $n - 1 = 3 - 1 = 2$, $\chi^2 = 9.21$ at 1%. Our value of χ^2 exceeds this. We therefore reject null hypothesis and conclude that there is good evidence that shape of bottle affects sales.

11.12 Null hypothesis: Poisson distribution with $m = 0.5$ is a good fit to the data. The total frequency is 100, thus expected frequencies are $100 \times P(r)$.

$$100 \times P(0) = 100 \times e^{-m} = 100 \times e^{-0.5} = 60.65$$

$$100 \times P(1) = 100 \times me^{-m} = 100 \times 0.5 \times e^{-0.5} = 30.33$$

$$100 \times P(2) = 100 \times \frac{m^2e^{-m}}{2} = 100 \times \frac{0.5^2 \times e^{-0.5}}{2} = 7.58$$

As the expected frequencies so far calculated total 98.56 and expected frequencies should exceed 5, there is no point in continuing.

O	E	O – E
63	60.65	+2.35
24	30.33	–6.33
13	9.02	+3.98
100	100.00	0

$$\chi^2 = \sum \frac{(O - E)^2}{E} = \frac{(2.35)^2}{60.65} + \frac{(-6.33)^2}{30.33} + \frac{(3.98)^2}{9.02}$$

$$= 0.091 + 1.321 + 1.756 = 3.168$$

Degrees of freedom = 3 − 1 = 2. (Note we did *not* calculate m; this was given.)

χ^2 (5%) = 5.99. Our value of χ^2 is less than this. We therefore accept the null hypothesis and conclude that a Poisson distribution with $m = 0.5$ is a good fit to the data.

Chapter 12

In Chapter 12 and in some of the model answers 'note form' has been used occasionally. This was done deliberately because students seem to remember the facts better if it is presented in this way. Some examiners will accept answers in 'note form' but most examiners will give good marks only if the answers are written in continuous prose.

12.1 a A simple random sample involves taking a sample of people from a population so that it is likely that the sample is representative of the population. Suppose that we wish to find information about employees of our company; we could obtain a list of names of all employees from the Personnel Department. We could then write the names of each employee on a separate piece of paper; we could then fold each piece of paper, place in a box and if we wanted a sample of 50 employees we would draw out 50 pieces of paper. The 50 employees so drawn would be a simple random sample.

Quota sampling involves taking a sample consisting of a fixed number of people (quota), usually by stopping people in the street. This is the method used by the B.B.C. to find out which radio and television programmes were heard or seen the previous day. To make certain that the sample represents the population, interviewers are given 'quota controls' to try to ensure that the sample is representative. For example, if the interviewer were asked to question 20 people, she (most interviewers are women) would be asked to interview 10 men and 10 women, people of all age groups and in all social classes. Quota sampling is quicker and cheaper than simple random sampling; however there is evidence that the sample is not as representative as simple random sampling. As interviewing is carried out in the street, people whose occupation takes them outside their firms' premises are more likely to be interviewed than people who do not leave their firms' premises during working hours.

b *Non-response.* If a survey involves interviewing a preselected list of people by calling at their houses, there are always some people on the list who cannot be interviewed. People can be out, be ill, moved away, refuse to be interviewed, etc. The inability to interview all the people on the selected list is called non-response. In a survey non-response can be a serious handicap because there is a possibility that those who do not respond are different from those that do; thus the results based simply on those that do respond would be biased. Steps must always be taken to reduce non-response (see page 201 for details).

c *Sampling frame.* In the discussion of simple random sampling in part (a) it was mentioned that to find a simple random sample of the company's employees it would be necessary to obtain from the Personnel Department a list of names of all employees. This list of names is called a sampling frame; thus a sampling frame is simply a listing of the population from which it is desired to draw a sample. In the U.K. there are two national sampling frames — the electoral register and the rating records, the latter being the records maintained by local authorities of premises in their area, the former the list of people eligible to vote in elections. For a list of the desirable properties of a sampling frame see page 194.

d *Multi-stage stratified sampling.* If a sample of 2 000 electors in Great Britain were required, a simple random sample would mean a spread-out enquiry; interviewers would have to travel long distances, particularly in country areas, just for one interview. A multi-stage sample of electors would aim to concentrate the interviewing in a number of small areas spread throughout the country. For example interviewing 20 people in a small area in each of 100 constituencies would simplify and reduce the cost of interviewing. The first stage of a multi-stage sample would be to choose 100 constituencies from all the constituencies in Great Britain, the second stage would be to choose at random a ward from all the wards in each selected constituency, and the third and final stage would be to take a random point on the register of electors for the ward and then to interview every 10th person on the electoral register until 20 names have been chosen. The effect of this procedure is to concentrate the interviewing in a few streets in each constituency that was selected at the first stage.

It is essential that the 100 constituencies are representative of all constituencies in Great Britain; a simple random sample could under-represent certain parts of the country or could include too many Conservative-held constituencies. To reduce this risk the standard procedure is to divide all the constituencies into strata in various ways — geographically England, Wales, Scotland, and regionally within each country; urban, rural, conurbation. A simple random sample is then drawn from each stratum.

12.2. a *Personal interviewing*

Advantages

i High response rate.
ii Answers to questions are spontaneous.
iii Questionnaire can be longer and more complex.
iv Interviewers can probe and explain the question to the respondent.
v Interviewers can observe and record observations (in sensitive areas such as racial origin the interviewer could enter on the recording schedule that a person was coloured without having to ask the question).
vi The interviewer can establish a rapport with the respondent and obtain answers to progressively more personal questions.

Disadvantages

i High cost of interviews (training, salaries and expenses).
ii Danger of interviewer bias (particularly at (iv) in advantages).
iii Possibility of response errors, e.g. entering the wrong answer or the respondent giving an answer which he/she thinks will please the interviewer.
iv Failure of interviewer to do the work properly e.g. completing all the questionnaires himself without bothering to interview the respondents (Supervisors have to be employed to check on interviewers)
v Because of cost a multi-stage design has to be used which has less precision.

Postal Questionnaires

Advantages

i Cheapness.
ii No danger of interviewer bias.
iii Can use a simple random sample or a stratified sample rather than a multi-stage sample and thus have higher precision.

iv Can have a larger sample and improve precision.
v Can obtain a considered reply.
vi The respondent can obtain information for replies by looking in his personal files or consulting other members of family.
vii Questions of an embarrassing nature can be asked.
viii Avoids the non-contact problem of interviewers; in the case of people who have moved the questionnaire is often forwarded to the addressee by the new occupants of the dwelling.

Disadvantages

i Response rates can be low. A response rate of over 80% has been achieved when professional bodies have conducted a survey of their members on a topic important to the profession (an interview survey using random samples is not likely to have a response rate much higher than 80%). In surveys conducted by academic researchers the response rates tend to be of the order of 50%. With commercial enquiries the response rates can be well below 50%.
ii Questions must be simple and capable of being understood by the least educated respondents.
iii Answers are not spontaneous.
iv As respondents tend to read all the questions before beginning to complete the questionnaire, answers are not therefore independent.
v No possibility of probing.
vi No evidence that the person to whom the questionnaire was addressed actually completed the questionnaire.

b When designing a questionnaire the following points should be considered:

i The number of questions should be as few as practicable.
ii Questions should be as simple as possible, the vocabulary should take into account the level of education of likely respondents, and technical terms should be avoided.
iii The use of closed questions where a number of alternative answers to the question are given should be used extensively. Provision should also be made for the respondent to be able to place a tick in a box against the answer he wishes to give. Open questions, where the respondent writes his answer, should be used sparingly, as the answers are difficult to code.
iv Questions should not be ambiguous.
v Leading questions should not be asked.

vi Avoid using vague words in questions, e.g. 'generally', 'fairly', 'roughly'.

vii If questions involving memory are asked, it must be recognised that people tend to remember only important events, e.g. a married woman would know how long she had been married, but possibly not how long ago she bought the dishwasher.

viii Questions should follow a logical sequence, e.g. ask if a person smokes before asking the number of cigarettes smoked per day.

ix Make provision for 'filter questions', e.g. if a person does not smoke give instructions to miss the question asking for number of cigarettes smoked.

x Ask easy questions at the beginning of the questionnaire, and the more difficult and more personal questions at the end of the questionnaire.

xi Make provision for coding the answers so that punch card operators can punch speedily from the questionnaire.

xii Check questions should be asked, e.g. ask if the respondents have a telephone, refrigerator or car, so that the percentage possessing these items in the achieved sample can be compared with National percentages.

xiii Test the questionnaire on a sample of people similar to those who will complete the questionnaire, and re-word questions which give difficulty.

c A sampling frame should have the following properties:

i *Adequate* — Does the electoral register cover the whole population of the survey? It excludes any person under the age of 16 years 8 months when compiled and excludes people not eligible to vote, mostly foreigners — this could be important in an area with a high concentration of foreigners.

ii *Complete* — The electoral register when compiled omits about 4% of those eligible to vote. This percentage varies with the age of voters and is particularly high for young voters, i.e. those between 16 years and 8 months and 18 years on compilation, where about 25% are omitted.

iii *Accurate* — The electoral register is compiled in October when it is 96% accurate; when published in February it is 93% accurate; and just before replacement, 85% accurate. The decay of the register varies from constituency to constituency; in certain London constituencies with a large population living in 'bed-sitters', the accuracy can drop to about 70% at the end of the life of the register.

iv *Free of duplication* — The electoral register does involve duplication (even though a person can vote only once) — people owning two houses, e.g. a London house and a country house, can appear on two registers; many students appear on two registers, one at their home address and the other at their college address.

v *Up to date* — The electoral register is compiled in October, is published in February and is not amended after publication. The electoral register includes people who have died or have moved away.

vi *Convenient* — The electoral register is readily available. It can be purchased. The British Museum has a complete set. The electoral register for a particular constituency is usually available in a central library in the constituency.

d See Question 12.4.

12.3 When considering which sampling method to use, a number of factors have to be examined. What precision is needed? What are the financial or resource constraints? Is a sampling frame available? Can the sampling frame be stratified? Is the population located in a small geographical area, or is the population spread nationwide? Is interviewing or a postal questionnaire to be used?

There are a number of alternatives to simple random sampling — stratified sampling, multi-stage sampling, multi-stage stratified sampling, systematic sampling, quota sampling. All these methods of sampling can be used in preference to simple random sampling in certain situations. If we have a survey where the population can be readily stratified and the sampling frame is also stratified and the strata differences are relevant for the survey, then stratified sampling is superior to a simple random sample. There is greater precision and furthermore information is available for each stratum. If a sampling frame is not available or cannot be compiled, then it is impossible to use simple random sampling; we would then have to use quota sampling or some other method of non-random sampling. If speed is essential, quota sampling is superior to simple random sampling. If we have a survey to be conducted by interview and the survey covers the whole country, then a simple random sample would be very expensive; in this situation a multi-stage stratified sample would be cheaper but would have lower precision when compared with a simple random sample of the same size. In this situation there is no doubt that a multi-stage stratified sample would be more cost effective.

A postal questionnaire can be sent to all parts of the U.K. for the same cost; in this case we would not use a multi-stage survey and a simple random sample would be preferred.

A simple random sample is easy to understand, and is easy to use, particularly in a small survey such as a school or college or in a small geographical area. It has the advantage that the standard error takes a very simple form $\frac{s}{\sqrt{n}}$. Thus in some situations simple random sampling is the best method to use, but to state that 'simple random sampling is the best method of sampling' is too simplistic. As outlined above, there are many situations when other methods of sampling are superior and there are occasions when simple random sampling is impracticable.

12.4 *Simple random sampling and quota sampling.* (See also 12.1(a).) A simple random sample would be used if the population were small or if a postal survey were being conducted where stratification of the population were impracticable. Quota sampling would be used if (i) a sampling frame were not available, (ii) speedy results are needed, (iii) costs are important — a quota sample costs less than half the cost of a simple random sample of the same size. The B.B.C. uses quota sampling for audience measurement because people needed to be contacted the following morning (otherwise the public forget what they saw or heard); random sampling takes longer to organise. Market research agencies use quota sampling extensively because of the cost factor.

Stratified sampling. This involves dividing the population into strata (the strata should be relevant to the survey, i.e. there should be strata differences) and then taking a sample from each stratum — the method of taking the sample could be simple random, systematic or quota. Stratification makes the sample more representative by ensuring that the correct fraction of each stratum is represented in the sample. In this way the standard error is reduced and precision improved. In multi-stage sampling stratification is frequently used to ensure that the first-stage process yields a representative sample of first-stage population units. In a college survey it may be desirable to stratify by departments and possibly by sex of students to ensure a representative sample.

Multi-stage cluster. (See also 12.1(d).) The multi-stage sample is used for national surveys. Firms organising opinion polls use this method. The Family Expenditure Survey and the General Household Survey both use multi-stage stratified sampling.

12.5 There are many things that can go wrong with a survey. A small pilot survey aims to detect major and minor defects of the survey. Any faults found by the pilot survey should be attended to before the main survey commences. The questionnaire is crucial to the success of any survey; the pilot survey should be used to detect the design faults listed in the answer to Question 12.2(b). In some pilot surveys several different questionnaires are used to try to find which form of wording to certain questions elicits the best response.

The pilot survey gives an indication of how long each interview takes, how long to make, say, 20 interviews, how many 'call-backs' should be made to obtain a reasonable response rate. In the case of a postal questionnaire, the effectiveness of various covering letters can be examined. The response rate, after the initial despatch of questionnaires and following reminder letters, and the costing of each response, whether by interview or postal questionnaire, can be established. It is possible to try in the pilot survey both interviewing and a postal questionnaire to see which is the better procedure, taking into account the relative costs and response rates.

Any organisational difficulties in the survey management, the sampling frame, the sample design, or instructions to interviewers, and degree of non response should be highlighted by the pilot survey.

It is essential to analyse the results of the pilot survey with care. Should the main survey actually proceed? Does the pilot survey indicate that the problem is too large for the resources available? Should the objectives be made more modest? Can the survey proceed with major or minor modifications? These questions need to be assessed and then a decision made.

12.6. The question states a 'limited budget'. If the budget is very small, then unless voluntary interviewers are available, there is no real alternative but to use a postal survey. If the budget is not too small, interviewers could be employed. As quota sampling is cheaper than random sampling, this would indicate that quota sampling should be used. If it were considered essential to be able to assess the accuracy of the estimates, then simple random sampling or a multi-stage sampling should be used.

A sampling frame would be necessary if a postal survey or random sampling were used. The electoral register would probably be the easiest to use; it is available (for most constituencies) at a central library and it would be a simple matter to select the sample at the library or to photocopy appropriate pages of the electoral register — it would be expensive to photocopy the whole register and as multi-stage sampling concentrates the interviewing in specific streets, it is cheaper just to photocopy the relevant streets.

Quota sampling entails setting up the quota controls. The population census returns for the county (available in the central library) in which the town is situated would enable you to find percentages in the occupational groupings and the age distribution, so that you can set up quota controls as shown on page 197. As the survey involves consumer durable goods which are mostly used by women, it may be desirable to question women only; if this is the case, then it is essential to have a quota control for women, e.g. working and non-working housewives.

When a decision has been made on which method is to be used in collecting the data (this is in effect controlled by the budget), a questionnaire or recording schedule must be designed (see 12.2(b)). With a limited budget a pilot survey probably could not be afforded, but as the survey involves obtaining factual information there should not be the same problems as in opinion surveys. The loss of the pilot survey should not be serious; nevertheless the questionnaire/recording schedule must be pre-tested to find drafting or other errors.

Whichever sampling method is employed, the sampling fraction — the ratio of sample size to the population — needs to be evaluated. The sample size of course is decided by financial considerations.

N.B. This question could have been tackled in a number of ways, for example stating initially that you would employ a postal questionnaire and discussing this method only. The model answer, by taking a 'general' approach, aimed to be more informative.

12.7 The corporation wish to interview 600 people out of a total of 60 000. This means a sampling fraction of $\frac{600}{60\,000} = \frac{1}{100}$

The total number of dwellings is 20 000 and with 60 000 inhabitants this means on average 3 persons per dwelling. If the contact is to be dwellings rather than individuals, this would mean 200 dwellings to be chosen; again the sampling fraction is $\frac{1}{100}$.

A decision must be made whether to interview 600 people or to select 200 dwellings and interview all occupants of each selected dwelling, thereby obtaining a sample of about 600 people. As a leisure centre would be used by young and old, we have to consider whether to use the electoral register or the rating record as the sampling frame. The electoral register does not include young people, therefore the rating record would probably be the better sampling frame and a sample of 200 dwellings the most suitable approach.

Stratified sample. The stratification factors should be relevant to the enquiry. We are not told whether tenure of dwelling and type of dwelling are relevant; however if we stratify by these factors, then we ensure that the sample is representative of the town. The ratio of privately-owned to corporation-owned is 8 000 to 12 000. We need to interview in 80 privately-owned dwellings and 120 corporation-owned, and as the sampling fraction is $\frac{1}{100}$ we should select this fraction of each type of dwelling. Thus for privately-owned dwellings we should select 20 town houses, 40 semi-detached, 10 detached 3 bedroom, 10 detached 4 or more bedrooms; for corporation-owned we should select 10 flats for the elderly, 10 semi-detached, 40 terraced houses, 20 low-rise flats, 40 high rise flats.

Multi-stage sample. This would aim to concentrate the interviewing and this would mean selecting some of the estates. There are 16 estates in total. The first stage sampling would be based on estates; we would select, say, 4 estates at random from the 16 estates. The next step would be to list the roads in the selected estates and choose roads at random and finally choose dwellings from the selected roads.

Multi-stage stratified sample. The stratification factors as above could be tenure of dwelling and type of dwelling. We could select, say, 2 estates from the corporation-owned estates and 2 estates from the privately-owned estates and to ensure that the sample is representative of the town we should select 40 dwellings per estate for each privately-owned estate and 60 dwellings per estate for each corporation-owned estate. For the selected estates it would be necessary to stratify by type of dwelling and choose the dwellings by a random procedure, but the sampling fraction for each stratum should be such as to ensure that the final sample for the whole town has the correct balance by type of dwelling.

As the council wishes to conduct the survey at minimum cost, then a multi-stage sample should be used, as this method concentrates the interviewing in small areas. Stratification helps to improve precision. Thus a multi-stage stratified sample would be the recommendation.

TABLE 1. THE NORMAL DISTRIBUTION FUNCTION

x	$\Phi(x)$		x	$\Phi(x)$		x	$\Phi(x)$		x	$\Phi(x)$		x	$\Phi(x)$	
0·00	0·5000	40	0·50	0·6915	35	1·00	0·8413	25	1·50	0·9332	13	2·00	0·97725	53
·01	·5040	40	·51	·6950	35	·01	·8438	23	·51	·9345	12	·01	·97778	53
·02	·5080	40	·52	·6985	34	·02	·8461	24	·52	·9357	13	·02	·97831	51
·03	·5120	40	·53	·7019	35	·03	·8485	23	·53	·9370	12	·03	·97882	50
·04	·5160	39	·54	·7054	34	·04	·8508	23	·54	·9382	12	·04	·97932	50
0·05	0·5199	40	0·55	0·7088	35	1·05	0·8531	23	1·55	0·9394	12	2·05	0·97982	48
·06	·5239	40	·56	·7123	34	·06	·8554	23	·56	·9406	12	·06	·98030	47
·07	·5279	40	·57	·7157	33	·07	·8577	22	·57	·9418	11	·07	·98077	47
·08	·5319	40	·58	·7190	34	·08	·8599	22	·58	·9429	12	·08	·98124	45
·09	·5359	39	·59	·7224	33	·09	·8621	22	·59	·9441	11	·09	·98169	45
0·10	0·5398	40	0·60	0·7257	34	1·10	0·8643	22	1·60	0·9452	11	2·10	0·98214	43
·11	·5438	40	·61	·7291	33	·11	·8665	21	·61	·9463	11	·11	·98257	43
·12	·5478	39	·62	·7324	33	·12	·8686	22	·62	·9474	10	·12	·98300	41
·13	·5517	40	·63	·7357	32	·13	·8708	21	·63	·9484	11	·13	·98341	41
·14	·5557	39	·64	·7389	33	·14	·8729	20	·64	·9495	10	·14	·98382	40
0·15	0·5596	40	0·65	0·7422	32	1·15	0·8749	21	1·65	0·9505	10	2·15	0·98422	39
·16	·5636	39	·66	·7454	32	·16	·8770	20	·66	·9515	10	·16	·98461	39
·17	·5675	39	·67	·7486	31	·17	·8790	20	·67	·9525	10	·17	·98500	37
·18	·5714	39	·68	·7517	32	·18	·8810	20	·68	·9535	10	·18	·98537	37
·19	·5753	40	·69	·7549	31	·19	·8830	19	·69	·9545	9	·19	·98574	36
0·20	0·5793	39	0·70	0·7580	31	1·20	0·8849	20	1·70	0·9554	10	2·20	0·98610	35
·21	·5832	39	·71	·7611	31	·21	·8869	19	·71	·9564	9	·21	·98645	34
·22	·5871	39	·72	·7642	31	·22	·8888	19	·72	·9573	9	·22	·98679	34
·23	·5910	38	·73	·7673	31	·23	·8907	18	·73	·9582	9	·23	·98713	32
·24	·5948	39	·74	·7704	30	·24	·8925	19	·74	·9591	8	·24	·98745	33
0·25	0·5987	39	0·75	0·7734	30	1·25	0·8944	18	1·75	0·9599	9	2·25	0·98778	31
·26	·6026	38	·76	·7764	30	·26	·8962	18	·76	·9608	8	·26	·98809	31
·27	·6064	39	·77	·7794	29	·27	·8980	17	·77	·9616	9	·27	·98840	30
·28	·6103	38	·78	·7823	29	·28	·8997	18	·78	·9625	8	·28	·98870	29
·29	·6141	38	·79	·7852	29	·29	·9015	17	·79	·9633	8	·29	·98899	29
0·30	0·6179	38	0·80	0·7881	29	1·30	0·9032	17	1·80	0·9641	8	2·30	0·98928	28
·31	·6217	38	·81	·7910	29	·31	·9049	17	·81	·9649	7	·31	·98956	27
·32	·6255	38	·82	·7939	28	·32	·9066	16	·82	·9656	8	·32	·98983	27
·33	·6293	38	·83	·7967	28	·33	·9082	17	·83	·9664	7	·33	·99010	26
·34	·6331	37	·84	·7995	28	·34	·9099	16	·84	·9671	7	·34	·99036	25
0·35	0·6368	38	0·85	0·8023	28	1·35	0·9115	16	1·85	0·9678	8	2·35	0·99061	25
·36	·6406	37	·86	·8051	27	·36	·9131	16	·86	·9686	7	·36	·99086	25
·37	·6443	37	·87	·8078	28	·37	·9147	15	·87	·9693	6	·37	·99111	23
·38	·6480	37	·88	·8106	27	·38	·9162	15	·88	·9699	7	·38	·99134	24
·39	·6517	37	·89	·8133	26	·39	·9177	15	·89	·9706	7	·39	·99158	22
0·40	0·6554	37	0·90	0·8159	27	1·40	0·9192	15	1·90	0·9713	6	2·40	0·99180	22
·41	·6591	37	·91	·8186	26	·41	·9207	15	·91	·9719	7	·41	·99202	22
·42	·6628	36	·92	·8212	26	·42	·9222	14	·92	·9726	6	·42	·99224	21
·43	·6664	36	·93	·8238	26	·43	·9236	15	·93	·9732	6	·43	·99245	21
·44	·6700	36	·94	·8264	25	·44	·9251	14	·94	·9738	6	·44	·99266	20
0·45	0·6736	36	0·95	0·8289	26	1·45	0·9265	14	1·95	0·9744	6	2·45	0·99286	19
·46	·6772	36	·96	·8315	25	·46	·9279	13	·96	·9750	6	·46	·99305	19
·47	·6808	36	·97	·8340	25	·47	·9292	14	·97	·9756	5	·47	·99324	19
·48	·6844	35	·98	·8365	24	·48	·9306	13	·98	·9761	6	·48	·99343	18
·49	·6879	36	·99	·8389	24	·49	·9319	13	·99	·9767	5	·49	·99361	18
0·50	0·6915		1·00	0·8413		1·50	0·9332		2·00	0·9772		2·50	0·99379	

(continued overleaf)

TABLE 1

x	$\Phi(x)$	Δ	x	$\Phi(x)$	Δ	x	$\Phi(x)$	Δ	x	$\phi(x)$	x	$\phi(x)$
2·50	0·99379	17	2·70	0·99653	11	2·90	0·99813	6	0·0	0·3989	2·0	0·0540
·51	·99396	17	·71	·99664	10	·91	·99819	6	0·1	·3970	2·1	·0440
·52	·99413	17	·72	·99674	9	·92	·99825	6	0·2	·3910	2·2	·0355
·53	·99430	16	·73	·99683	10	·93	·99831	5	0·3	·3814	2·3	·0283
·54	·99446	15	·74	·99693	9	·94	·99836	5	0·4	·3683	2·4	·0224
2·55	0·99461	16	2·75	0·99702	9	2·95	0·99841	5	0·5	0·3521	2·5	0·0175
·56	·99477	15	·76	·99711	9	·96	·99846	5	0·6	·3332	2·6	·0136
·57	·99492	14	·77	·99720	8	·97	·99851	5	0·7	·3123	2·7	·0104
·58	·99506	14	·78	·99728	8	·98	·99856	5	0·8	·2897	2·8	·0079
·59	·99520	14	·79	·99736	8	·99	·99861	4	0·9	·2661	2·9	·0060
2·60	0·99534	13	2·80	0·99744	8	3·0	0·99865	38	1·0	0·2420	3·0	0·0044
·61	·99547	13	·81	·99752	8	3·1	·99903	28	1·1	·2179	3·1	·0033
·62	·99560	13	·82	·99760	7	3·2	·99931	21	1·2	·1942	3·2	·0024
·63	·99573	12	·83	·99767	7	3·3	·99952	14	1·3	·1714	3·3	·0017
·64	·99585	13	·84	·99774	7	3·4	·99966	11	1·4	·1497	3·4	·0012
2·65	0·99598	11	2·85	0·99781	7	3·5	0·99977	7	1·5	0·1295	3·5	0·0009
·66	·99609	12	·86	·99788	7	3·6	·99984	5	1·6	·1109	3·6	·0006
·67	·99621	11	·87	·99795	6	3·7	·99989	4	1·7	·0940	3·7	·0004
·68	·99632	11	·88	·99801	6	3·8	·99993	2	1·8	·0790	3·8	·0003
·69	·99643	10	·89	·99807	6	3·9	·99995	2	1·9	·0656	3·9	·0002
2·70	0·99653		2·90	0·99813		4·0	0·99997		2·0	0·0540	4·0	0·0001

The function tabulated is $\Phi(x) = \frac{1}{\sqrt{2\pi}}\int_{-\infty}^{x} e^{-\frac{1}{2}t^2}dt$. $\Phi(x)$ is the probability that a random variable, normally distributed with zero mean and unit variance, will be less than x. The last two columns give the ordinate $\phi(x) = \frac{1}{\sqrt{2\pi}}e^{-\frac{1}{2}x^2}$ of the normal frequency curve.

Index